NET DESTRUCTION

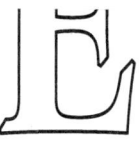

The Death of Atlantic Canada's Fishery

Kent Blades

NIMBUS PUBLISHING LTD

Copyright © Kent Blades, 1995

All rights reserved. No part of this book covered by the copyrights hereon may be reproduced or used in any form or by any means—graphic, electronic, or mechanical—without the prior written permission of the publisher. Any request for the photocopying, recording, taping, or information storage and retrieval systems of any part of this book shall be directed in writing to the Canadian Reprography Collective, 379 Adelaide Street West, Suite M1, Toronto, Ontario, M5V 1S5.

Nimbus Publishing Limited
P.O. Box 9301, Station A
Halifax, NS B3K 5N5
(902) 455-4286

Design: Arthur B. Carter
Cover photos: Department of Fisheries and Oceans

Canadian Cataloguing in Publication Data
Blades, Kent.
Net destruction
Includes bibliographic references.
ISBN 1-55109-097-X

1. Fisheries—Atlantic Provinces. 2. Fishery policy—Canada.
3. Sustainable fisheries—Atlantic Provinces. I. Title.

SH224.A8B52 1995   639.2'09715   C95-950275-0

Dedicated to the loving memory of my father, Clifford A. Blades, a hard-working family man, who was freed upon reaching retirement age to pursue his dream of life on the water as a groundfish fisherman.

I would like to thank several people for their assistance and support in the completion of this project. First, thanks go to Dorothy Blythe and Paula Sarson at Nimbus Publishing for managing this project and seeing me through its delays and revisions. Thanks to Carl Myers, Joe Gough, and Norwood Whynot at the Department of Fisheries and Oceans Communications Branch in Halifax for their help with retrieving background reports and providing transparencies included in the book. A special thank you to Timothy Nickerson of North West Harbour, who acted as my technical advisor, ensuring that this landlubber did not go astray when relating details of life and work on the sea. Finally, loving thanks go to my wife, Cathy, and three sons, Joshua, Matthew, and Andrew, who often had to put up with a grumpy bear over the past two years as my dream of writing this book became a reality.

Contents

Introduction IV

Chapter 1:
From Boom to Bust: The Atlantic Groundfishery in Crisis 1

Chapter 2:
Life Underwater: Habits and Habitats of Atlantic Groundfish 18

Chapter 3:
Blaming Nature: In Search of Environmental Scapegoats 38

Chapter 4:
Widening the Divide: Who Should Be Allowed to Fish for What 47

Chapter 5:
Players for Profit in the Atlantic Groundfishery:
Offshore Giants Cast Long Shadows 83

Chapter 6:
False Data, False Ethics: Impact of Government Policies
on the Fisheries 104

Chapter 7:
Politics of Alliance: Collusion and Coercion Among Government,
Big Business, and Scientists 121

Chapter 8:
Quotas and the Food Chain: Will Partial Restrictions Work? 138

Chapter 9:
Fish or Fishermen First: Tobin's Ecosystem Approach to Fisheries 158

Chapter 10:
Running Out of Time: Blueprint for a Sustainable Groundfishery 170

Bibliography 180

Introduction

The world's fisheries are at a crisis. Over the last decade, the majority of stocks fished commercially around the globe have been depleted at an alarming rate. Gone is that long-held notion that the seven seas harbour a secure, bountiful, and limitless resource. We have flexed our technological muscle, unwittingly imposing our will upon the oceans' delicate supporting ecosystems, only to expose an unforeseen vulnerability. The relentless pursuit of fish to satisfy the needs of a hungry world has proven that the treasures of the seas are not inexhaustible.

The groundfish industry of Atlantic Canada has not escaped this global decline in resource. By July 1992, when the Department of Fisheries and Oceans (DFO) finally announced a northern cod moratorium, the damage had already been done. The reduced quotas, which had been gradually introduced for offshore and inshore sectors since the late 1980s had not been effective; fish stocks continued to decline. Thousands of people in hundreds of communities dotting the Atlantic coastline from Labrador to southern New Brunswick were hit with the impact of declining groundfish stocks and the final moratorium. Scores of people were thrown out of work as plants closed and millions of dollars worth of fishing boats and gear remained tied up at the docks. Those who have traditionally made their living from the sea are used to cycles of boom and bust, but this crisis is much more serious than its predecessors. This time it is not only the economic viability of industry players that is in question but the long-term survival of the resource itself, while a time-honoured way of life hangs in the balance.

This book is written for those who can regrettably relate to the present fishery crisis experienced by Atlantic Canadians. Although extensive in its scope, this account is not intended to be an academic study or a research document. These are better left to those who write from the halls of academia.

At the same time, it is not a polite narrative or a casual commentary on how the Atlantic groundfishery crisis came about. Rather, this book is intended as an exposé—an attempt to reveal the real reasons why the fishery is in such a crisis now. It examines why the Atlantic fishing industry is at risk and why the future of our fishing communities and their families is in jeopardy. The book explores how the crisis arose through the roles played by the various stakeholders. Culpability is noted where it is due. Ultimately, this book is meant to inform opinions—perhaps change them—and to chart a new course for the Atlantic groundfishery. It is meant to give voice to the unheard or silenced Atlantic Canadians concerned for their future and to stir those in positions of authority to a speedy resolution of the crisis.

There are hundreds of Atlantic coastal communities that are experiencing the horrendous effects of the fishery crisis. Clark's Harbour is one of them, a small town of just over one thousand residents, located on Cape Sable Island at the southern-most tip of Nova Scotia. In many ways it is typical of the more than four hundred Atlantic Canada coastal communities that have been devastated by the crisis. First occupied by the Mi'kmaq and later by the French and English, Clark's Harbour's roots have always been firmly planted in the sea. From the mid-1680s, when European colonists first took up the challenge of establishing permanent homes on the rocky landscape, the community focus has been on the fishery. The early settlers soon realized that the poor soil and rocky terrain precluded the success of farming or lumbering. Their survival depended solely on the bounty of the adjoining sea that would honour their faith and reward their hard work.

Over the years, the fishermen and industry leaders of Clark's Harbour and Cape Sable Island have proven themselves to be a progressive lot, ready to accept technology and technological advances, adopting ideas that would improve productivity. They were masters during the age of sail, yet they readily and rapidly embraced the transition to motor-driven vessels. Experience and experimentation developed the hook and line method of fishing, later enhanced by the introduction during the 1870s of the revolutionary trawl, which utilized mile-long anchored lines set with thousands of hooks. The most noteworthy adaptation, however, came with the development of a sturdier, more seaworthy fishing vessel—the now world-famous Cape Islander that originated on Cape Sable Island.

The local boatbuilding industry at Clark's Harbour began as a family affair. Developed by Ephriam Atkinson of Clark's Harbour at the turn of the twentieth century, successive generations have continued the modifications and improvements demanded by industry. The Cape Islander proved popular among Atlantic coast fishermen from Newfoundland to the New England states. With rapid acceptance of this design, the local boatbuilding industry steadily grew to mirror the established confidence of those involved in the groundfish and lobster industries. By the early 1980s more than twenty local boatshops were turning out jet-aged fibreglass versions of the Cape Islander. The boatbuilding industry had become an integral part of the local economy, providing high-paying jobs and profitable returns for proprietors.

Similar to most progressive fishing communities, Clark's Harbour underwent a major transformation after World War II. Along with the improvements in vessels and gear used in the fishery to increase catch sizes, processing plants became much more efficient with the use of forklifts, processing machines, and refrigeration. These upgrades were made feasible by the development of a North American market for fresh and frozen groundfish fillets. By the late 1960s, the local fishing industry was booming. British Columbia Packers had established a large processing facility in Clark's Harbour. Sable Fish Packers, located about 5 km (3 mi.) beyond the town limits, underwent a major transformation and expansion to take advantage of the developing prospects. Both longline and dragger fishermen were enlisted to supply the plants. Despite frequent catches in excess of 45,000 kg (100,000 lbs), the local fleet was unable to adequately meet the growing demand. By the late 1970s, fish was being trucked in from as far away as Cape Breton Island to meet market commitments.

The entire community was caught up in the boom. Young men had always been susceptible to the call of the sea and a chance to fish with their fathers, but by the late 1970s through the mid-1980s, the call was irresistible. It was an exceptional time for Atlantic Canada's fishery. For many, there was good money to be made and school would have to wait. With the growing need for plant workers, many women joined the paying workforce for the first time. Times were good and consumer confidence grew. New homes were built and sporty cars and half-tonne trucks lined paved driveways.

The euphoria continued until the mid-1980s. These were good times not only for the groundfish industry but also for the lucrative lobster industry that continued its recent upward climb, fetching record returns for area fishermen. The swordfish and Bluefin tuna fisheries attracted increasing attention from fishermen, while the herring and herring roe fisheries swelled the ranks of fish plant workers, albeit only for the summer months. Even with the workforce supplemented by homemakers and school-aged children, seasonal shortages of plant workers existed. As a result, there was a significant influx of workers from other regions of Nova Scotia and from as far away as Newfoundland.

In the midst of this prosperity, some local visionaries warned of a pending groundfish decline and industry downturn, but few dared to contemplate the disaster about to befall Clark's Harbour and communities like it throughout the Atlantic region. Groundfish catches began their free fall in 1986 and 1987, though the effect on the industry and local economy was masked somewhat by strong market prices. Fish plant workers were still able to work sufficient weeks to qualify for Unemployment Insurance, so family incomes were sheltered during the first two years of the downturn. It was not long, however, before the effects of reduced groundfish quotas made an impact. Fishermen were being squeezed at a time when they could least afford it because they were unable to fish long enough to qualify for the same levels of Unemployment Insurance. Many had recently invested heavily in expensive new boats and gear, banking on a continued bright future. By the late 1980s, inshore fishermen were having to fish harder and further from shore.

At the same time, the boatbuilding industry was also brought to its knees with strict government enforcement of vessel replacement guidelines that were designed to halt any further expansion in catching capacity. Boatshop after boatshop completed its last vessel and closed its doors. Fish plant after fish plant scaled back their operations to reflect a shrinking supply of landed product. The former British Columbia Packers plant in Clark's Harbour, which had employed as many as three hundred workers in the prosperous years of the 1970s and 1980s, closed in 1989. The Sable Fish Packers operation was effectively reduced to dependence on the processing of herring and herring roe. A town and a community, so typical of the many rural isolated fishing villages dotting the Atlantic coast, had become increasingly dependent upon the lobster industry and federal income assistance programs. The latter involved three possible options: income support; re-training for either a professional approach to the fishery or work outside the industry; or early retirement packages. Re-training in particular was underfunded and oversubscribed.

Today the people of Clark's Harbour remain numbed and shaken by the swelling tide of events. The financial strain of being unable to provide a steady and adequate income has placed tremendous stress upon breadwinners and their families. Where once a fisherman set out before dawn to spend long days setting and hauling his fishing gear—now the same fisherman spends his time mending his gear and checking his equipment without knowing for sure if he will have the opportunity to use them. A drive around Cape Sable Island shows that housing construction has slowed to a crawl and that more and more family homes are being put up for sale. Parents who hoped to send their children away for a university education are having to reconsider those plans.

The young people of the area, faced with fewer and fewer employment opportunities, are being forced to leave for Ontario and western Canada. Local and provincial governments openly admit that they can do little to create employment for those displaced workers and young people located beyond commuting distance of the metropolitan areas, where more job prospects exist. Although the federal and provincial governments have made attempts to compensate for job losses and to establish re-training programs, such as The

Atlantic Groundfish Adjustment Strategy (TAGS), discussed in Chapter 9, fishing communities realize that the measures are only damage control; they don't guarantee a future.

Indeed, one has to wonder what future is left for communities like Clark's Harbour. Will they be left for ghost towns as government support programs become exhausted and standby fisheries, such as lobster, are depleted? Will those who may hold some of the answers to recovery passively watch, or will they find a way to get their message across? Will those in authority learn from past errors in judgement and technique and commit themselves to the measures necessary to restore and protect groundfish stocks in order to breathe life back into our coastal communities? Can we sufficiently temper greed for profit with conservation for the sake of the future? The following chapters explore the contributing causes to the crisis and suggest measures to salvage the Atlantic groundfishery. If we can agree to exercise prudence in the fisheries, perhaps an eventual rejuvenation will be possible.

Chapter One

From Boom to Bust: The Atlantic Groundfishery in Crisis

For nearly a century, the Atlantic groundfishery has been one of the economic mainstays of Atlantic Canada. The rich fishing grounds of Georges Bank and the Grand Banks have supported a growing industry from the early 1800s, when the salt bank schooners first began their brisk trade in salt cod, to the modern fast-frozen fishery of the 1950s. Over the years, generations of families have made their living in the fishery, and although markets and stocks have fluctuated from time to time, the incomes of fishermen and industry workers gradually improved over the years. But since the early 1990s, employment in the groundfishery has become an uncertain and destitute occupation for many people whose families have lived and worked for generations in the region's coastal communities. Those employed in the groundfishery were accustomed to dealing with cycles of boom and bust, but over the past five years the industry has had to cope with cumulative lows never before experienced in catches and quotas. Federal biologists employed by the Department of Fisheries and Oceans continue to report that groundfish stocks are still declining with no sign of stabilizing. Stocks of some of the most popular groundfish species, such as cod and haddock, have been reduced to a level from which experts are concerned they may never rebound. The fear is that their low numbers may be insufficient to permit the replenishing of stocks and the continuation of the species.

While some dare to talk of a future with a sustainable groundfish resource, evidence dating from the early 1990s indicates that the cumulative effects of steadily increasing harvest, combined with unusually harsh environmental conditions, have placed the continuation of the groundfish resource in peril. On an international level, the 1992 report of the reputable World Resources Institute (WRI) indicates that global catches were then in decline and would continue to slide. The WRI pointed to statistics provided by the Food and Agriculture Organization (FAO) of the United Nations in 1991: "...Most traditional marine fish stocks have reached full exploitation; that is, an intensified fishing effort is unlikely to produce an increase in catch, and the use of any new fishing methods that did increase the catch would cause overfishing and, therefore, a further decline in fish populations. Four of FAO's seventeen major marine fishing areas are overfished." In fact, hindsight suggests that the increased catches of the 1960s and 1970s can be attributed to the increased effort to exploit the resource. The increased catches were not the result of a growth in fish stocks but an expansion of fishing effort.

The turn of events, from boom to bust, is not surprising. In many ways the exploitation of world fish stocks mirrors that of the northwest Atlantic groundfish and, for that matter, other natural resources. Invariably, humans have overexploited the earth's resources, often to the point of extinction. An examination of the history and development of the Atlantic coast groundfishery clearly illustrates this human bent.

Although people have fished—for food, for profit, and for recreation—development of world fisheries proceeded at a leisurely pace until the Industrial Revolution, which swept Europe and the Western Hemisphere during the nineteenth century. Once the pace of "progress" picked up, there was no turning back. Continual technological improvements in fish capture and in fish culture developed over the past fifty years have resulted in a quintuplicating of commercial production, with a new world record set in 1993—101 million tonnes produced internationally. This recent record would seem to contradict the claims of a fishery crisis.

However, the sources of that increase raise serious concerns. Recent increases in world production have come almost entirely from aquaculture (underwater agriculture), which now accounts for 16 million tonnes of the world harvest. In addition, a growing portion of the international take captured at sea (about 30 per cent) now consists of small pelagic fish. These pelagic species, which frequent the surface layers of the ocean, are of lower commercial value and are used, almost exclusively, in the industrial production of fertilizers and fish meal and oils. Production from pelagic stocks masks the decline in catch of the preferred white-fleshed groundfish stocks such as cod, haddock, and flounder that established their dominance during the nineteenth century and maintained their high rank until the 1960s. A build-up of excess catching capacity of groundfish by world fishing powers meant that stocks in both the northeast and northwest Atlantic and the north Pacific were eventually fished out. Consequently, fleets were forced to look farther afield for catches and to consider other species as alternatives to those preferred by consumers.

The development of the Atlantic Canadian groundfishery parallels that of the global fishery; it, too, began at a leisurely pace. The first peoples to inhabit what has become known as Atlantic Canada were the Mi'kmaq, Beothuk, and Malecite nations. They were here an estimated ten thousand years before the Europeans established a foothold in the seventeenth century. Fish formed an important part of the subsistence lifestyle of these aboriginal peoples. Each season brought its bounty, as documented by Father Pierre Biard, a Jesuit missionary, who lived on the southern coast of Nova Scotia from 1611-1616. His writing is quoted in Wells's *The Fishery of Prince Edward Island.*

"...In January they have the seal hunting (seal oil was regarded as a great delicacy and stored for use as a sauce).... In the month of February and until the middle of March, is the great hunt for beavers, otters, moose, bears, caribou.... In the middle of March, fish begin to spawn ... often so abundantly that everything swarms with them. After the smelt comes the herring at the end of April; and at the same time bastards (Canada geese), ... sturgeon, and salmon, and the great search through the Islets for (waterfowl) eggs.... From the month of May up to the middle of September, they are free from all anxiety about their food; for the cod are upon the coast, and all kinds of fish and shellfish.... (In September) the eels spawn.... In October and November comes the second hunt for elks and beavers and then in December ... comes a fish called by them *ponomo* (probably Tommycod) which spawn under the ice."

The aboriginal peoples gathered shellfish along the shore at low tide, fished with crude hooks and lines, speared fish, and prepared traps (including weirs) built across the mouths of rivers and at the shore's edge. Fishing was an integral part of their culture and way of life. The fishery of those early days posed little threat to fish populations, save for the few instances where a lake or bay may have come under intense pressure from an expanding human population.

While native peoples were content with a seasonal harvest of nature's provisions, European explorers, navigating the world's oceans in search of treasures and fortune, were immediately taken with the promise of profit in the waters of the northwest Atlantic. In 1497, John Cabot, a Venetian sailor under exploratory charter issued by the British Crown, visited what are now the Grand Banks off Newfoundland. There he found an untapped fish resource literally bubbling to the surface. So plentiful were the cod and haddock that his crew needed only to lower baskets over the side of their small craft and haul the ocean's bounty on board. The commercial possibilities of the New World fish were immediately evident. The British Empire would no longer have need of the resource supplied by Iceland from which great supplies of fish were imported annually. A limitless and bountiful resource had been found for the benefit of a fledgling empire.

What John Cabot and his crew had discovered was a sampling of the early groundfish resource of the northwest Atlantic. Groundfish are so called because they generally feed and live near the seafloor. Although there are many species that fall under the groundfish category, the most significant ones in terms of economic importance include Atlantic cod, haddock, pollock,

Net Destruction

redfish, halibut, Greenland halibut (turbot), white hake, silver hake, American plaice, yellowtail flounder, witch flounder, and winter flounder. Groundfish of lesser economic importance include tomcod, cusk, catfish, and skate.

The groundfishery, recognized early on for its economic potential, quickly became the foundation of the Atlantic fishery. The abundance of groundfish and their versatility as a staple in the human diet has determined their popularity. Groundfish, with cod at the pinnacle, are generally white-fleshed and comparatively low in fat, making them particularly suitable for salting and sun curing as well as for processing into a fresh or frozen product. It was this former method of preservation (i.e., salting) that proved of vital importance in the development of the early northwest Atlantic—salt-cured cod was often considered the medium of exchange, or currency.

The extension of the land mass of Atlantic Canada known as the continental shelf offers near ideal habitat for groundfish because of ideal water temperature and ample food supply. Distinct plateaus of shallower water on the shelf, known as banks, are especially attractive to various species of groundfish and serve to isolate stocks within that species from others of their kind. It is these banks that have proven to be the focus of the groundfishery of Atlantic Canada for abundance of stocks.

Many of the commercial species of the northwest Atlantic are pictured here, including groundfish, crustaceans, and molluscs.

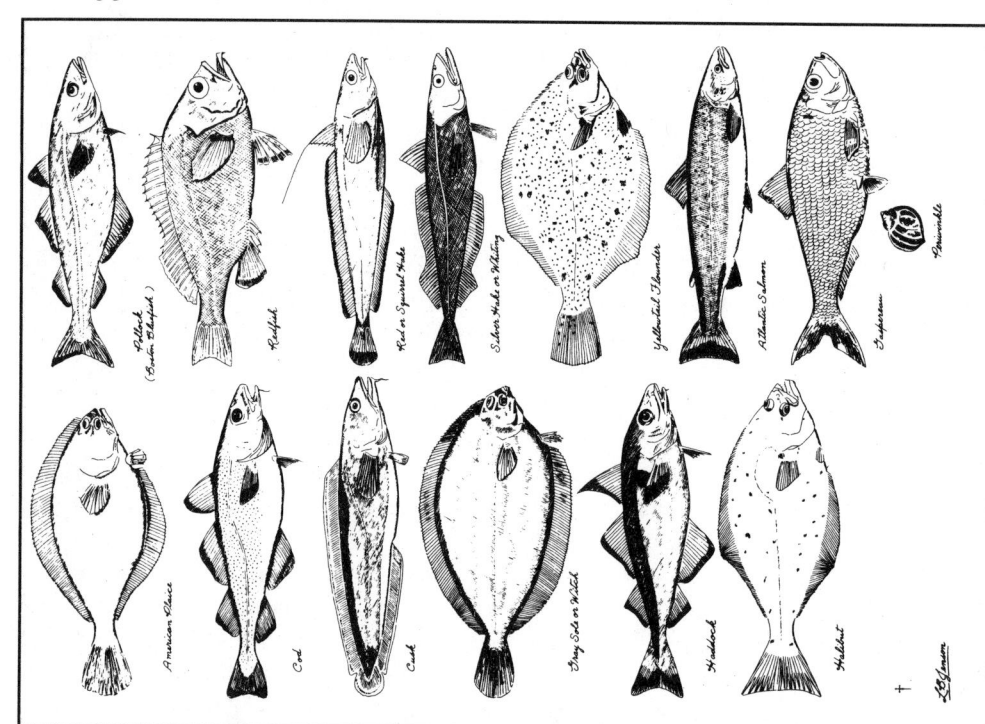

Cod has, without dispute, been the most important commercial species fished in the northwest Atlantic. This kingpin of the groundfish ranges from the shallower inshore waters of about 5 m (16 ft.) in depth to the edge of the continental shelf in waters as deep as 600 m (1,968 ft.). There are twelve identified stocks within the Canadian zone from Frobisher Bay in the north to Georges Bank in the south. Each stock of cod is a biologically distinct unit and, though similar in appearance and behaviour to other geographically located stocks, shows little tendency to intermingle with adjacent members of the species. Each stock is identified by the geographical region in which it is located and includes the following: Northern Labrador; Southern Labrador-Southern Grand Bank; Flemish Cap; Southern Grand Bank; Bank St. Pierre; North and East Gulf of St. Lawrence; Sydney Bight; Southern Gulf; Banquereau-Sable Island; Browns Bank; and Georges Bank.

Haddock is the choice of fish-and-chip lovers. This flavourful, white-fleshed species was, at one time, found in commercial abundance in Canadian waters from the southern coast of Newfoundland south to Georges Bank. Six stocks were identified including Georges, Browns, and Sable Island banks. Haddock was, at one time, abundant with three stocks fished commercially: Placentia Bay, Bank St. Pierre, and the southwestern area of the Grand Bank, all off the southern coast of Newfoundland. Commercial popularity led to depletion of these latter three stocks.

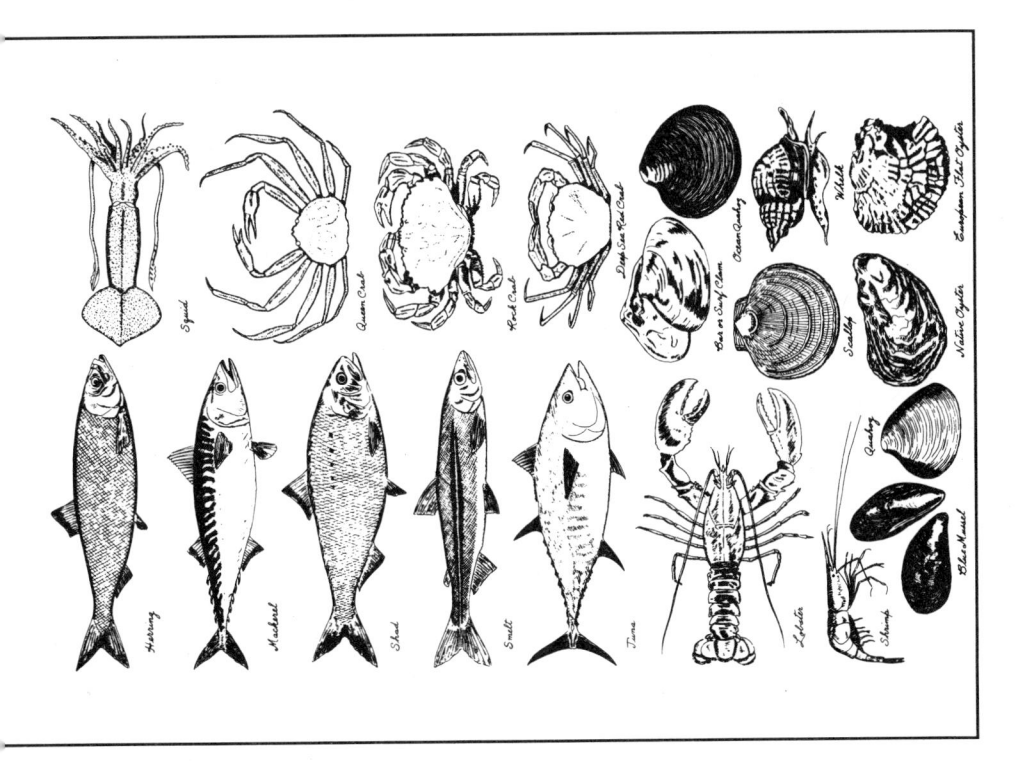

Pollock is the third member of the commercially important triumvirate and is similar in appearance to the other two—cod and haddock. Much less is known about pollock than about cod and haddock, though it is generally assumed that only one stock exists and that its range is from the Bay of Fundy to the waters off eastern Cape Breton.

Redfish, also known as ocean perch, has gained importance with the demise of species such as haddock in the fresh and frozen fillet trade. This species prefers the deep cold waters of the continental shelf from Labrador south to the Gulf of Maine. Though little is known about redfish stock definition, commercial fisheries for this species were developed for the Flemish Cap and Gulf of St. Lawrence.

Atlantic halibut is another prized groundfish species favoured for its fresh-fish taste. Halibut, a flatfish, ranges from southern Labrador to the Gulf of Maine. The principal catch areas include the Scotian Shelf, Gulf of St. Lawrence, and Grand Banks of Newfoundland.

Turbot, or Greenland halibut, is preferred in the fresh or frozen fillet trade. The species is found in Canadian waters from the Arctic south to Georges Bank. Traditionally, this species has been fished from the deeper waters near the coasts of Newfoundland, Labrador, and Baffin Island, and in the Gulf of St. Lawrence.

White hake has been used locally as a fresh-fish commodity, though for export it requires salt curing. This groundfish species is most commonly found in the Gulf of St. Lawrence, on the western Grand Banks, and on the Scotian Shelf from Cape Breton to Georges Bank.

Silver hake is a slender, bony fish considered underutilized by the domestic fishery. This fish deteriorates quickly unless frozen soon after capture and has, therefore, played a minor role in the Canadian Atlantic fishery. Foreign fleets have, however, fished the species in the deep waters near Browns and Sable Island banks. The flesh of silver hake is used in the production of minced fish products.

Included in the many flatfish species distributed from Baffin Island to the Gulf of Maine are American plaice, yellowtail flounder, witch flounder, and winter flounder. These species are filleted and prepared as a fresh or frozen product. American plaice is the most common of the four species with five stocks identified, though the area of greatest abundance is the Grand Banks. The yellowtail flounder is found throughout the Canadian Atlantic though, again, the area of primary distribution and abundance is the Grand Banks of Newfoundland. Winter flounder is found and fished throughout the continental shelf waters of Atlantic Canada, while the witch flounder frequents the deeper waters of the Gulf of St. Lawrence and the waters off the coasts of Labrador and Newfoundland.

The ragged Atlantic coastline provides numerous sheltered bays and harbours adjacent to these fish species and fishing grounds. As a result, hundreds of fishing villages sprang up early on to take advantage of the vast marine resources found at their doorsteps. The French were the first to set up camp, establishing a colony on the Bay of Fundy in 1604. They were soon followed by the British. The groundfishery was the resource base and the raison d'etre for communities which steadily grew in number through the eighteenth and nineteenth centuries. By the early eighteenth century, some forty-five hundred fisherfolk and farmers populated the coast of Nova Scotia, the most preferred district in the region. The number of communities established in the Atlantic region approached one thousand by the end of the nineteenth century and included those from what is now Labrador to the Bay of Fundy and the American border. Their names became synonymous with the fishing industry and included such towns and villages as Trepassey and Burgeo in Newfoundland; Canso and Lunenburg in Nova Scotia; Caraquet in New Brunswick; and Souris in Prince Edward Island.

Little definitive statistical information is available regarding the exact size of the groundfish catches or the profits of the fishery through the early years. It is known that prior to 1550, the cod fishery of Newfoundland involved the annual visit of 128 fishing vessels sailing from European ports. The fishery continued to expand all along the Atlantic coast after 1550, and by the late 1600s the annual Newfoundland catch of cod had reached almost 100,000 tonnes. By the late 1700s the amount of cod landed had approached 200,000 tonnes, annually. Through the 1800s the cod catch for the entire Atlantic region showed considerable variation, ranging from a low of 150,000 tonnes to a high of approximately 400,000 tonnes per year.

The 1800s were heralded as the golden age for Atlantic Canada. In New Brunswick, three generations of lumber barons and lumberjacks played double roles as shipbuilders and sailors. The clipper ships constructed at the busy shipyards of Nova Scotia formed one of the world's leading merchant fleets. All the while the fisheries continued to be one of British North America's most valued assets.

Increases in the catch rate were brought about as the Atlantic fishery welcomed two major improvements in capture technology during the mid-1800s—the cod trap in Newfoundland and the "bultow" in Nova Scotia. The Newfoundland inshore fishery adopted the cod trap in the 1860s and immediately improved their catches. The cod trap is shaped like a box and open at the top. It is constructed with curtains of knitted twine positioned to lead the cod into the box and discourage their escape. Prior to the arrival of the cod trap, groundfish had been captured primarily by the use of towed seines, or nets, or they had been caught on single-baited hook and lines.

Nova Scotia fishermen were quick to embrace a French innovation called the "bultow," or trawl method of fishing, made popular during the 1870s. This mode involved the utilization of a mile-long anchored line that was set with thousands of baited hooks. Fishermen were, for the first time, able to take advantage of the great stocks of groundfish which were found on the offshore grounds. Sailing to the banks in schooners equipped with small wooden boats called "dories" from which a trawl was set, crews were able to return home from the Grand Banks with catches in excess of 440,000 kg (200,000 lbs) of salted fish.

By 1911 there were twenty-two hundred boats, primarily foreign vessels, participating in the offshore cod fishery on the Grand Banks and grounds of the Scotian Shelf off Nova Scotia. A spring and summer cod fishery also developed on the east coast of Labrador with vessels sailing from Newfoundland and Nova Scotia, and from as far away as New England to take part in the seasonal fishery which had to await the break-up of saltwater ice floes each year. With the introduction of the stern trawler and the otter trawl method of fishing—both of which drag massive nets just above the seafloor—and with more effort directed toward groundfish, the Atlantic Canadian catch increased to weigh in at approximately 900,000 tonnes, annually, by the middle of the twentieth century.

The steady though uneven development of the Atlantic Canada groundfishery through the twentieth century increased community dependence upon the harvest of the sea. Many fishermen had traditionally worked both the land and the sea, supplementing their income with part-time farming or work in the woods. With more and more emphasis being placed on the commercial fishery, families had less time to practise self-sufficiency, gradually giving way to a decline in the home-based production of food and other goods and a greater dependence on store-bought items. By the middle of the twentieth century, many households had sacrificed much of their independence in favour of increased reliance upon the commercial fishery.

Despite increasing dependence on the groundfishery and the improved efficiency in catching fish, independent fishermen on the whole did not become wealthy at this time. In fact, fishermen's incomes, with few exceptions, remained below Canada's official rural poverty line. The economic situation was even worse for households of part-time fishermen. However, few options existed to allow or to encourage fishermen and their families to leave the industry. Changing professions would require relocating, the breaking of strong family ties, and often those from rural fishing communities had limited education or training, since many left school at a young age to work in the fishing industry.

With domestic offshore development in the groundfishery emphasized since 1977, an Atlantic processing sector emerged that seemed destined to assume a position of importance and power in communities. In many isolated fishing communities, a single processor not only purchased and processed the fish landed but typically offered for trade or sale the fishing supplies and bait required for the next trip to the grounds. In some local villages, the loop of commercial enterprise was further closed—the processor often doubled as general merchant. A relationship of dependence and mutual benefit naturally developed within most fishing communities between fishermen and processors.

The larger, less isolated fishing communities most often offered a broader-based commercial sector. Frequently, a number of independently owned fish plants competed for the allegiance of fishermen. In that process, an entire web of support industries sprung up to serve the groundfishery, including boatbuilding, equipment supply, and the transportation of goods. The importance of the fishery was magnified by its links to other sectors, such as shipbuilding and repair, fishing supply and manufacturing equipment, transportation and distribution of product and supplies, and provision of a host of personal and business services ranging from legal to consulting work. In addition, incomes earned directly from the fishery or indirectly through the service sector were spent both locally and regionally on consumer goods and services which, in turn, employed a broader spectrum of the workforce.

Although the Atlantic Canada groundfishery continued on an even keel through the first half of the twentieth century, the stocks of the North Sea and northeast Atlantic began showing signs of ill health prior to World War I. The development of both wind- and steam-powered trawlers by 1900 had opened up the Atlantic coastal waters to intense fishing. North Sea stocks of cod, haddock, and plaice were targeted by world fleets, and by 1910 the average catch per vessel started to show signs of decline and the catch composition was altered. The prized flatfish species, such as sole and turbot, were soon fished down, so attention shifted to plaice and haddock. Despite an increase in effort by fleets working the North Sea, the overall groundfish yields showed a noticeable decline. A strong recovery of stocks during World War I, when commercial fishing was effectively scuttled, confirms the declines were the result of increased fishing effort.

Although the respite afforded by world conflict permitted a rebound of stock health, a further slide was soon precipitated by gradual improvements on a worldwide level in fish catching power through the 1920s. By the end of that decade, the traditional North Sea grounds were showing the effects of overfishing, and larger vessels in the European fleet turned their attention to waters to the north and west. After 1930, the distant water fleets intensified their search for cod and groundfish in the northwest Atlantic. Using huge vessels equipped to fish the rough seas year-round, in 1939 the Europeans recorded a catch of nearly 1 million tonnes of groundfish in the Barents Sea north of Scandinavia and the former U.S.S.R. The catch of cod in waters near Iceland peaked at a 1/2 million tonnes in 1933. It had become increasingly obvious that the aggressive European fleet would have to look further afield to satisfy its growing appetite for fish.

Dramatic improvements in fish catching capability on the world stage, fostered by the stepped-up industrialization following World War II, were destined to have an impact on the leisurely pace enjoyed until then by the Atlantic

Canadian fishery. The traditional salt cod fleets employed by Spain, Portugal, France, and Italy were joined off Canada's east coast in the late 1950s by the new generation of long-distance freezer trawlers from the former U.S.S.R., Poland, Japan, Cuba, and the former East and West Germany. In 1953 total catches by foreign fleets of northwest Atlantic cod weighed in at 900,000 tonnes. By 1968 the foreign fleet catch had more than doubled to a peak of 1,900,000 tonnes. In 1950 only one-third of the catch was fished by foreign fleets; this proportion doubled to nearly two-thirds at the peak of the fishery in 1968. At the same time, Canada's share of the total groundfish catch tumbled because it could not compete with the larger capacity foreign vessels.

The Canadian government's response to the aggressive dominance by distant water fleets was to compete head-on with the newcomers by subsidizing the development of a domestic offshore. The traditional inshore sector fishermen who relied on traps, gillnets, handlines, and trawl were left largely to their own devices. As a result, the number of Canadian offshore vessels, weighing over 45 tonnes (the original classification is 50 tons) and fishing the northwest Atlantic, increased from 211 in 1959 to 558 in 1968—an increase of 367 vessels in just nine years. Subsidy programs implemented by the Canadian federal government also encouraged the production of larger vessels, which led to a dramatic increase in catching capacity.

The growth of the Canadian offshore sector further aggravated a worsening situation for inshore fishermen. With Canadian jurisdiction extending only 19 km (12 mi.) to sea prior to 1977, the inshore sector often had to contend with large foreign vessels fishing off their doorstep and often on the same grounds they had fished for years. Their own government unwittingly increased that pressure by sponsoring a build-up of the domestic offshore. The result was a dramatic drop in inshore catches across the entire Atlantic region. The inshore northern cod fishery of Newfoundland, for instance, witnessed steady catches of approximately 180,000 tonnes annually through to 1959, but with the marked increase in take by the combined domestic and foreign offshore, that figure dropped to 35,000 tonnes in 1974.

The intense fishing pressure exerted by the distant water fleets was not isolated to traditionally fished stocks, such as cod on the Grand Banks. The cod stocks off the coast of Labrador were also targeted. As well, underutilized species such as silver hake, shark, and tuna were targeted by the Soviets, Japanese, and Faeroese. Cod catches by foreign fleets from waters off the coast of Labrador climbed from 60,000 tonnes in 1959 to more than 250,000 tonnes in 1961. The silver hake catch by foreign fleets off Nova Scotia increased dramatically, showing a one-year jump from 9,000 tonnes in 1962 to 123,000 tonnes in 1963.

The international industry was not prepared for the sudden crash of groundfish catches that occurred after the peak year of 1968. In retrospect, the outcome of such a build-up in fishing power seems entirely predictable, although few believed that government would let it happen. The newly created Canadian offshore, in league with the distant water fleets, had all but wiped out one of the world's abundant natural resources. Few believed it was possible to exhaust the treasures of the sea, even with the latest technological advances. It was unthinkable that the groundfish stocks of the northwest Atlantic could be overfished. But by the early 1970s the cod catch off Newfoundland for domestic and foreign fleets had been reduced to a mere 200,000 tonnes; the Grand Banks cod catch was registered at less than 30,000 tonnes.

A similar fate befell the haddock stocks on Georges Bank, off the southern coast of Nova Scotia. The stock had gained some semblance of health, after it had been nearly wiped out by the push during the late 1920s to satisfy the newly developed American frozen fillet market. However, the haddock stock was ravaged by foreign vessels during their 1960-1970 invasion, with the result that it was again threatened with annihilation.

The distant water fleets had not only plundered the stocks of the northwest Atlantic but had raised a worldwide level of concern by extending their influence throughout the seven seas. In a push for a more competitive groundfishery, Canada felt justified in taking the action demanded by its domestic fishing industry to extend its territory. In 1977 Canada unilaterally declared the extension of its territorial waters to a limit of 332 km (200 mi.) from shore—hence, the 200-mile limit. It was hoped that the problem with foreign overfishing would become a thing of the past, a nightmare properly exorcised. The promise was there—Canadians wished to claim control of the entire seabed along the Atlantic coast and put an end to the pillage attributed to the foreign fleets.

Some Canadians considered Ottawa's initiative a bold step, while to others it seemed insufficient. Ottawa appeared content with only partial victory, yielding to the need for a neat line to be drawn by cartographers. The 200-mile limit paid no recognition to the fact that Grand Banks stocks, important to Newfoundland and Nova Scotia fishermen, would now straddle that boundary. In their natural yearly cycle, northern cod stocks frequent the deeper water on the Nose and Tail of the Grand Banks and Flemish Cap, outside the line where they are still exposed to intense pressure by the foreign fleets searching for product. The original intent was to have jurisdiction extend beyond 200 miles, where necessary, to encompass the entire continental shelf. Instead, Ottawa surrendered its position, indicating that it did not care to make waves by including all of the shelf as nature would have dictated.

Negotiations during the two-year period leading up to extended Canadian jurisdiction were international in scope; therefore, it was appropriate that these duties fell primarily to those in the federal Department of External Affairs. Given that this department, with offices located far from the Atlantic coast, had other concerns, including trade in manufactured goods from politically powerful Ontario and Quebec, as well as other resource items to exchange, it

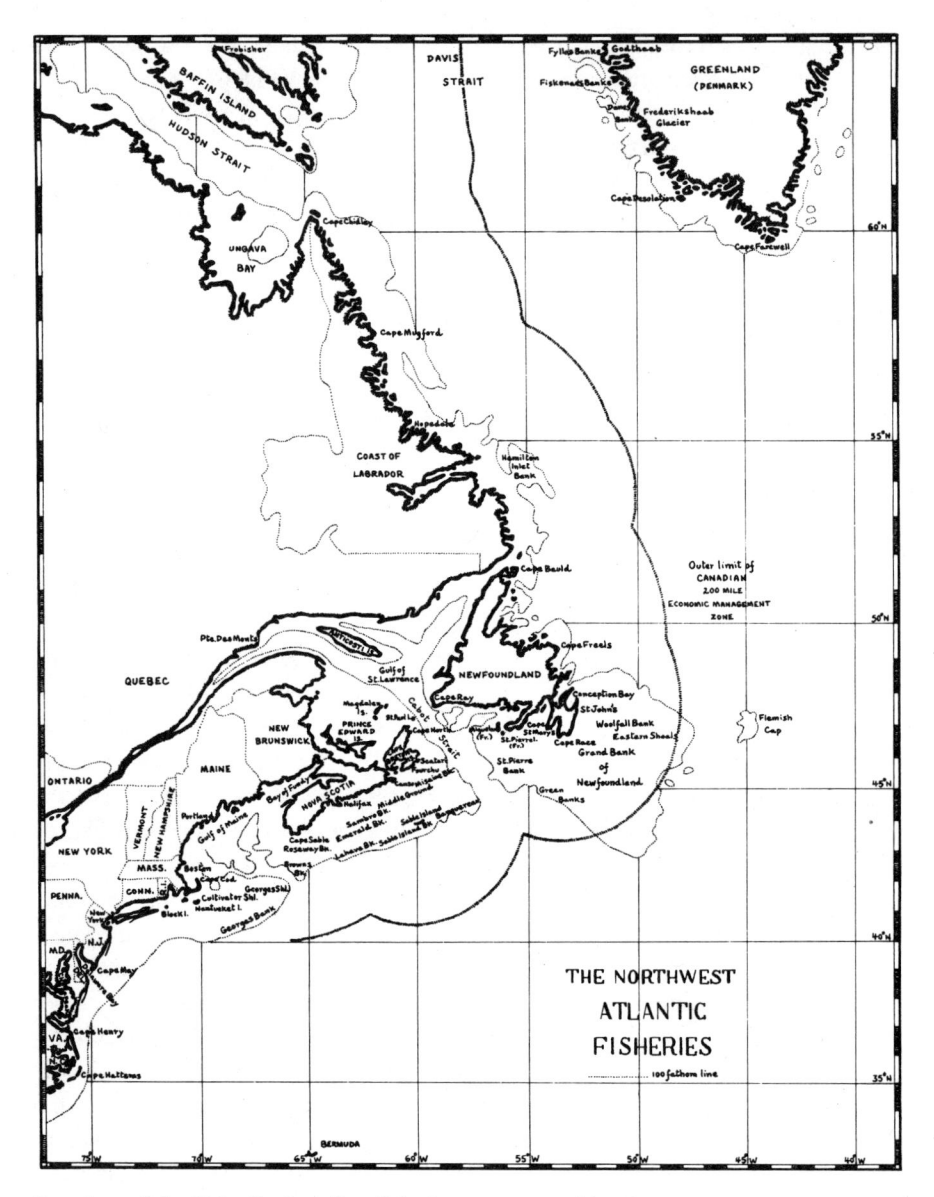

Boundary of the 200-mile Canadian fisheries zone, created in 1977.

was only natural that a less than perfect deal was made for the fishing industry. This was a decision that would haunt successive Canadian governments having to cope with recurrent groundfish crises and the subsequent conflicts from foreign fleets fishing stocks that straddle the Canadian boundary.

What was truly astonishing was that Canada fumbled its primary mandate. With the foreigners pushed outside the 200-mile limit, the federal government had a prime opportunity to establish a management strategy that would ensure a sustainable groundfish fishery, at least for the near future. The yearly quotas and their related regulations should have been set to provide ample raw product for the domestic industry while, at the same time, leaving more than enough in the water to protect stock health from any contingency. Rather than taking a cautious and conservative approach to the development of the fishery, the federal government encouraged further expansion of both the domestic offshore and inshore catching and processing capacity. The aim was to position the national fishery to capitalize on the expected bonanza of increased catches. The industry, buoyed by the enthusiasm of politicians and bureaucrats, set aside their inhibitions and invested heavily in the tools of expansion.

The anticipated offshore success precipitated a rapid expansion by certain large fishing firms in Nova Scotia and Newfoundland in the late 1970s. The directors of these companies had equated catching and processing capacity with marketing ability. The latter, however, was their downfall. By the early 1980s the region's largest fishing firms found themselves in a financial crisis that arose from unrealistic expectations. High-priced, unsaleable fish inventories were part of the problem, but the crisis was compounded by high interest rates, a relatively strong Canadian currency that undermined the industry's export competitiveness, and good-size fishing catches by competitor nations such as Iceland.

The Canadian government, fearing that its experiment with industry building would be a failure without supportive intervention, stepped in to bail out the large debt-ridden corporations. Government's solution was to shoulder much of the financial burden, write off a portion of the debt, and institute a new quota system called "Enterprise Allocations" (EAs) for the offshore sector. The result was that the federal government married Newfoundland's bankrupt companies under the family name of Fishery Products International (FPI). Meanwhile, in Nova Scotia, an infusion of new, private funds and a generous handout of taxpayers' money returned National Sea Products (NSP) to a stable footing. Both of these companies began to benefit from the generous grant of EAs—an annual quota of groundfish to be taken when and as the company harvest managers directed.

However, another opportunity to guarantee a bright future for the fishery was squandered. Government had failed to scale back the catching and processing capacity of these large offshore corporations to more realistically reflect a level necessary to allow for a sustainable harvest of groundfish. The wisdom of quotas for a sustainable fishery was not immediately evident. In fact, it took another six years for the necessary prudence to be fully recognized.

After enduring year-end losses in 1984 and 1985, NSP ran a string of three successive years in the black. FPI fared even better, buying back the shares government had claimed during the restructuring in 1983. By 1987, ten years after the declaration of the 200-mile limit, both the present and the future appeared to be in good hands. Nova Scotia's industry was credited with emerging leaner and meaner from the crisis of the early 1980s. In 1986 the factors which had once undermined Atlantic Canada's fishing effort appeared to be working in its favour. Canadian groundfish were commanding premium prices on the Boston market, and in some cases double the 1985 prices. The weak Canadian dollar, combined with the decline of foreign competition, had helped the region's industry increase to record highs the value of exports to the crucial American market.

Despite the persistent troubles and uncertainties of the past, the fishery as a whole remained hopeful. The number of workers in the industry climbed steadily, until about 1990. The number of registered fishermen in Atlantic Canada continued to grow, and dependence upon the fishery increased. The number of registered fishermen totalled 43,500 in 1978. It grew to 59,000 in 1988, and to approximately 64,000 in 1990, despite the gathering storm clouds of crisis. Similar increases have characterized employment in the processing sectors. The number of fish plant workers grew rapidly after 1977, and despite a dip in 1983 (as a result of large-company restructuring) the numbers increased to 45,000 workers in 1988, and further to 60,000 workers in 1990.

All the while something very disturbing was occurring. Despite very rosy predictions of sustained Canadian catches, the Newfoundland inshore in particular was troubled by the fact that the northern cod were not returning to the harbours and bays of the rocky province, as they had in the past. The cod were getting smaller in size each year and more gear had to be set just to keep up with minimum expectations for catches. In 1984 the inshore fishermen were complaining to the DFO that stocks were being taken by the large and powerful offshore fleets before they could make their seasonal migration to the inshore grounds. The federal government politely listened but failed to take effective action.

The decline in catches and the increasing amount of work to sustain catch levels for Newfoundland inshore fishermen failed to improve over the next five years. They were joined in the chorus of despair by the Newfoundland offshore sector in 1989. Northern cod, once hailed as the promise of abundance, had failed to live up to its billing as saviour of the Atlantic groundfishery. A 1980 study prepared for the Economic Council of Canada had predicted good times for the Atlantic industry and a harvest of 365,000 tonnes for 1985. Actual landings for that year fell far short of the prediction, totalling a mere 195,000 tonnes. In 1988, eleven years after extended Canadian jurisdiction, the catch peaked at only 245,000 tonnes. From there it was all downhill. Catches totalled 133,000 tonnes in 1991 and a minuscule 21,000 tonnes the following year, when the fishing season was shortened by the imposition of a northern cod moratorium at midyear.

Since 1977, on the Scotian Shelf off Nova Scotia and in the Gulf of St. Lawrence off the western coast of Newfoundland, a powerful inshore dragger sector was also at work. Encouraged by the promise of an abundance of fish and by incentives to increase catching and processing capacity, this sector had an aggressive beginning after the extension of the territorial limit. By the mid-1980s individual fishermen and small processors had built new, more powerful vessels and plants capable of handling four times more fish than were actually available.

Groundfish stocks off Nova Scotia, including cod, haddock, pollock, and flounder, had been in difficulty since the early 1980s, but the jump in market prices in 1985 and 1986, fostered by increased demand from health-conscious consumers, distracted attention from the advancing predicament. For reasons unknown to government scientists, Scotian Shelf stocks showed more resilience than their northern cod cousins to the north, despite being fished at perhaps two to three times the prescribed rate. Perhaps this resilience is inherent in the stock, or maybe the warmer waters on the Scotian Shelf are responsible for their enhanced fecundity and improved recruitment (that is, a measure of the number of individuals that join the fish of legal harvestable size). In any case, these Scotian Shelf stocks showed themselves better able to withstand the increased pressure from both the domestic inshore and offshore sectors. Quotas for all fleets were reduced but the fishery off southwestern Nova Scotia was allowed to hobble along. The industry in this quarter, accustomed to going full throttle, had to adjust to the cutbacks, though it did not have to endure a complete closure.

Facing reduced quotas and falling fish prices, the Atlantic groundfish industry was forced to make some tough decisions in 1989 and 1990. A number of processing companies, including both NSP and FPI announced plant closures and reductions in operations throughout the Atlantic region, with several thousand fishermen and plant workers losing their jobs. NSP closed four plants in Nova Scotia and three more in Newfoundland. By 1993 the flagship of the Atlantic Canada groundfishery business, National Sea, was left with only three plants in operation. This included the main plant in Lunenburg, Nova Scotia; a part-time operation in Arnold's Cove, Newfoundland; and a plant south of the border in Portsmouth, New Hampshire. The latter two plants were processing fish caught in the north Pacific by American and Russian interests. The Lunenburg plant, which had customarily operated year-round, was shut down for a good portion of the summer.

Reduced quotas for the inshore sector were felt all along the Atlantic coast. Particularly hard hit was the inshore processing sector of Newfoundland, hindered for some time by a shortage of raw product. The quota reductions forced a number of closures among small plants throughout the Atlantic region. Some operations were more fortunate, however. This was especially the case for those clustered in the southwestern region of Nova Scotia. Many of these operations benefited from their involvement in a diversified fishery, which included lobster, herring roe, and tuna.

The news that most everyone came to expect but no one was prepared for was announced on July 2, 1992, at the Radisson Hotel in St. John's. Federal Fisheries Minister John Crosbie chose the same hotel to announce a moratorium on the fishery for northern cod that he had used three years earlier to dismiss any idea that a closure of the fishery would become necessary. Minister Crosbie had failed to heed earlier advice given by his department's scientists that northern cod stocks were in eminent danger of collapse and should receive immediate protection. The 1992 announcement meant that thirty thousand Newfoundlanders would be put out of work—the shock waves were felt throughout the region. This was the news event that brought the groundfish crisis to national and international attention simultaneously. The news grew worse. Eighteen months later a new government and a new Minister of Fisheries and Oceans made another announcement. Brian Tobin, taking the hard line in favour of conservation of the groundfish stocks, closed all but one Atlantic Canada cod fishery (that off southwestern Nova Scotia) and severely cut quotas of all other species, putting a further five thousand people out of work in January 1994.

Fishing industry workers throughout the Atlantic region joined Newfoundlanders in their shocked response. How could they feed their families on the $211-$382 weekly compensation package made available, by application, from the federal government to fishermen and plant workers with a history in the groundfishery? For the entire decade of the 1980s, the Atlantic coast fish business was profitable for those involved. Fishermen and those in the fish processing sector had worked long hard days, taking advantage of large catches and strong prices. The good years had been quickly translated into a visible increase in the standard of living for many coastal communities; it was a decade of gold rush mentality. Although few Atlantic Canada fishermen had become wealthy, the returns often afforded the highliners and more successful fishermen gains in material wealth, all of which were threatened by cuts to quotas and fisheries closures.

As for who is responsible for the present groundfish crisis, some would point to the lack of or ineffectiveness of government policies, others to the widespread greed of big companies and of fishermen, and most to some gear sector other than their own. Some actually place the blame for the fishery crisis on the greed and self-interest of all players. "We are all guilty," some industry workers astutely claim.

Malcolm Nickerson, age sixty, of Lower Clark's Harbour, Cape Sable Island, like many able-bodied fishermen along the coast, would like to be free to return to the fishing that he had and his fathers before him had—working from dawn till dusk. His ancestors were fishermen, and they were generally rewarded in proportion to their efforts and skill. Today, however, there is despair as the crisis in the groundfishery takes its toll on him, his family, and his community.

It is only early September 1994, and this once-proud fisherman, like so many others, is already overhauling his lobster gear in preparation for the upcoming season scheduled to begin some four months down the road. He is simply putting in time, all the while trying to convince himself that the repairs are best completed early. In years past, he would have taken every advantage of the later summer weather, charging ahead to ensure a good portion of his annual income. Ordinarily, he would still be on the water either handlining or setting tubs of baited trawl for groundfish. Instead, he is on shore, in a sombre, reflective mood.

"Over the past few years fish have gotten scarcer and scarcer," he laments. Fewer and fewer fish have found their way to fishermen's hooks. It was in the early 1980s that Malcolm and fishermen from southwestern Nova Scotia noticed cod and other groundfish were becoming smaller, and they had to work harder to make the same return. While the demise of the groundfish stocks did not occur overnight, the decline was predictable and predicted. The pending tragedy was lamented by fishermen long before it made headline news by a probing mass media. Its pain was first felt as a persistent knot in the pit of the stomach of many fishermen in tune with the demands of the ocean's complex and delicate balance. Too many fish were being removed too rapidly from the marine food chains, and there appeared no way to avert certain disaster.

While some would argue that fish stocks naturally undergo cycles of decline and recovery, influenced largely by uncontrollable environmental factors, few would deny that the impact of modern day fishing has become more deadly, placing increased pressure on groundfish stocks. In an attempt to curb fishing effort and protect marine resources, management schemes applied by coastal nations, including Canada, have failed miserably to account for the increased capacity of modern fleets to locate, chase, and kill. What has happened to fish stocks around the world, and specifically to the Atlantic groundfishery, must be recognized as the inevitable result of an exploitative mind-set—the belief that what is modern, more efficient, and faster is also better. And to be better is a necessity if exploiters of resources are to compete in a fast-moving global economy.

Where did it all go wrong that Atlantic Canada and world fisheries are now faced with such gloomy prospects? Will new management measures and adjustment programs put the industry back on course or simply prolong the heartbreak and misery of communities in decline? Will revised measures lead to improvements in the health of prized groundfish stocks? Or is a sustainable groundfishery for Atlantic Canada even possible at this late point in the crisis? And what will become of Atlantic Canada's coastal communities?

Chapter Two

Life Underwater: Habits and Habitats of Atlantic Groundfish

Sea water covers more than two-thirds of the earth and shelters a wondrous underwater world. This web of life below the waves has thrived for eons, but it has become painfully obvious that human intrusion has taken a toll. What are the definitive reasons, then, that commercially fished stocks, specifically northwest Atlantic groundfish, have suffered a prolonged, and perhaps irreversible, period of decline? Is it simply a result of the increased take of targeted species or, more troublesome, are modern harvesting methods altering the physical and biological features of the ocean strata? Perhaps our ocean's systems are far more delicate than previously imagined. Possibly the ocean's habitats are being altered by human encroachment, rendering them less hospitable and perhaps incapable of supporting marine life. To better understand the complex factors underlying the present groundfish crisis, we must examine what is presently known about their habits and habitats.

Even after four hundred years of harvesting groundfish in the northwest Atlantic, there is much we do not know about the world under the ocean. It is a complex system beyond the comprehension of most landlubbers and a formidable challenge for those scientists and naturalists striving to unravel the mysteries of this dynamic environment, a system which has provided a source of wealth and enterprise for thousands of coastal communities.

In spite of the persistent mystery, some fundamental knowledge has been gleaned about the static and dynamic forces at work in the sea. These forces determine where fish species will gather and live out their lives. Bottom formations, for example, have been the subject of much recent study. New information, aided by modern computerized technology, has allowed for more detailed charting of the thalassic, or ocean, terrain. Raw data from surface surveys can now be translated overnight into full colour, three-dimensional, virtual reality graphics, enabling researchers to scan large segments of the ocean floor and to more accurately determine its physical features. An inventory of data about prevailing tides, currents, and sea water flow will potentially enable scientists to detect changes in these physical characteristics. The changes could then be correlated with changes in the health of fish stocks, permitting more accurate predictions about the impact of harvesting methods on the health and habitats of fish.

The terrain of the seafloor is as varied as the land, with ridges, troughs, mountains, plateaus, and basins. It is quite unlike the flat ocean floor envisioned by early scholars. Directly offshore from the coastal plains are wide, shallow-water bottom-lands called continental shelves. These areas are the supporting ecosystems for commercial fish stocks. The shelf surface features a gentle slope and is relatively smooth, except where it has been cut by submarine canyons. Some canyons extend seaward from the mouths of rivers, like the Laurentian Channel, between Cape Breton and Newfoundland, that is carved by the flow from the St. Lawrence River. Most canyons and channels are related to the action of ancient streams that were active when the ocean surface level was much lower. The Fundian Channel, which separates Georges Bank from Browns Bank off the southwestern tip of Nova Scotia, is thought to have been sculptured by the flow of such an ancient stream. Beyond the outer reaches of the shelves is the continental slope, which descends steeply to the deep ocean flatlands or basins, once believed to be largely barren. These basins are also marked by deep-water mountain ranges and canyons.

Just as with land soils, the bottom mud of the continental shelves is deposited in distinguishable layers. The upper portion is generally full of life and aerobic, while further down, conditions are more anaerobic, limiting the spectrum of life forms. Disruptions to the continental shelf environment, resulting in the mixing of layers, change its character. Not only are attached and burrowing life forms physically disturbed, but their micro habitat is altered, endangering established spatial arrangements. On the other hand, disruptions to these soils allow the vital release of nutrients, which are occasionally projected upward where they are used by life forms floating or swimming on the surface. Alternately, when plants and animals die, they fall from the upper portions of the water column, nourishing life on the seafloor.

While the continental shelves are generally smooth and without marked relief, there are portions which are characterized by peaks and troughs. The rugosity of the ocean bottom is important for marine life because it helps determine water current speed. Increased rugosity reduces current speed in some areas of the floor, which increases the suitability of the local environment for attached and free-swimming life forms. Other areas benefit from the effects of increased current speed because of a gain in nutrient exchange.

The dynamic forces at work, shaping conditions for growth of plant and animal life beneath the sea, involve the vital interplay of tides, currents, and surface winds together with the intrusion of external water masses. The

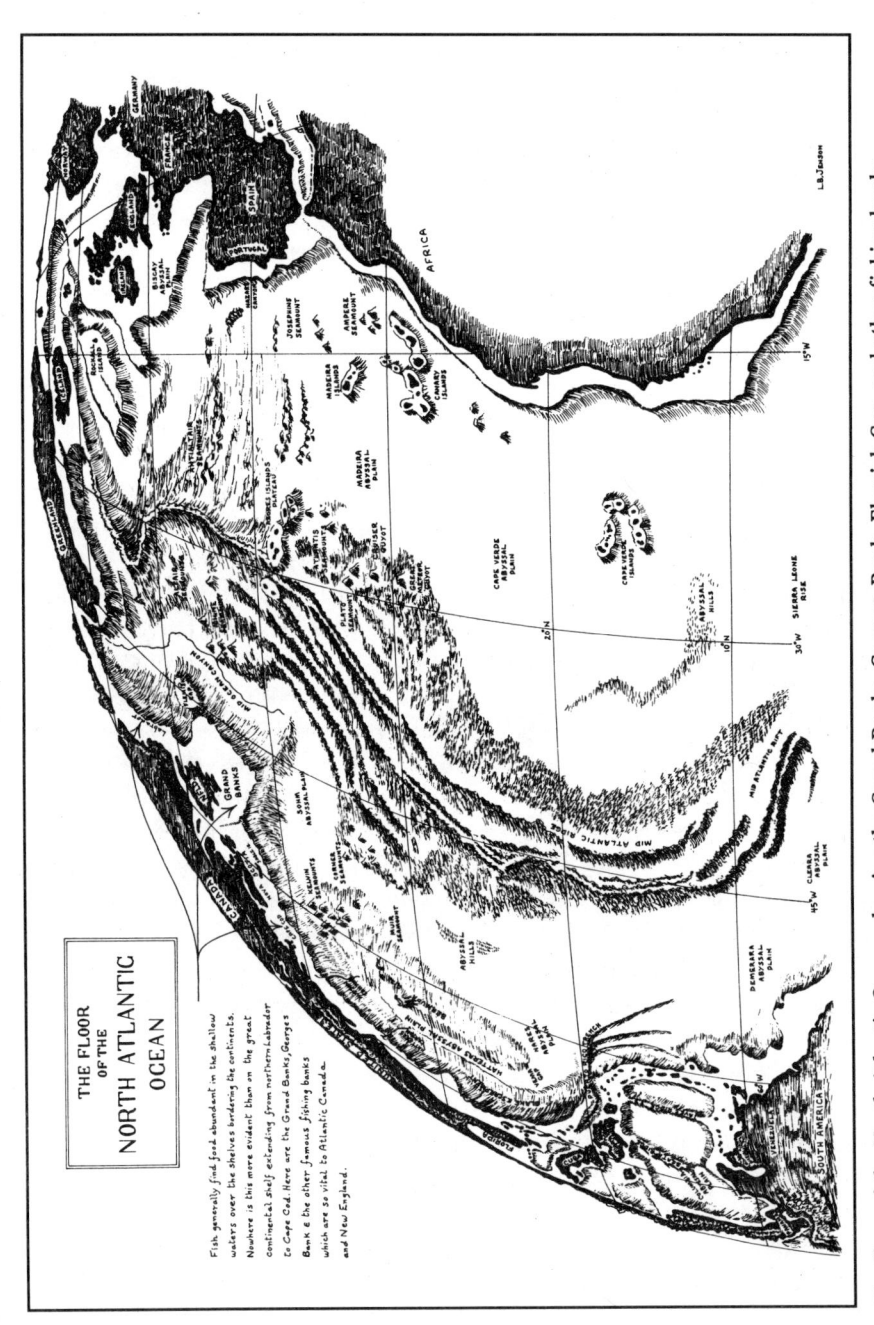

The floor of the North Atlantic Ocean, showing the Grand Banks, Georges Bank, Flemish Cap, and other fishing banks where groundfish stocks were abundant. The map indicates the marked relief of the seafloor terrain.

dynamic forces influence the static ones, such as bottom soil type and seafloor relief. These physical elements determine the location of fertile areas in the sea, that is, areas capable of supporting lower life forms in the food chain which support those valued species further up.

Ocean tides are the regular rise and fall of sea level caused by the gravitational pull of the sun and moon. Twice a day waters accumulate and rise along the shore, creating a high tide. Each high tide is followed by an ebb. The tides of the Bay of Fundy are world renowned, exhibiting a difference of 15 m (50 ft.) or more between high and low tide. Significantly less variance in tide levels is experienced in the Gulf of St. Lawrence (less than 2 m, or 7 ft.) and along the remainder of the Atlantic Canadian coast. Tides play a major role in the exchange of nutrients in marine habitats.

Ocean currents, less visible than tides, are rivers of sea water that flow through the upper layers of the water column. Some currents run as deep as 18 m to 30 m (60 ft. to 100 ft.), and they can vary in speed from 3 km (2 mi.) a day to 32 km (20 mi.) a day through the Cabot Strait and the Bay of Fundy. Ocean currents also play a significant role in the transport of nutrients along the Atlantic coast. The forces driving them are the differentials of temperature they exhibit, either warmer or colder than the waters through which they pass. The primary ocean currents are the warm North and South Equatorial Currents, which flow from east to west along the equator in each of the oceans. The equatorial currents split when they reach the continents. The North Equatorial Current changes course, turning north along the east coasts of continents in the northern hemisphere and deflecting its warm waters into high latitudes. The Gulf Stream off the east coast of North America, flowing from its origin in the Gulf of Mexico, is an example of this movement. It slows in velocity to a widening drift off Newfoundland, splitting as it flows eastward across the Atlantic. Reaching waters off the British Isles and Europe, it exerts a moderating influence on what would otherwise be a harsh and cold climate, thereby also offering favourable temperatures for groundfish spawning.

A lesser known element influencing water mixing and nutrient availability is the influx of warm water masses from the continental slope. This phenomenon has been identified as playing a role in the productivity of waters off southwestern Nova Scotia and at the mouth of the Bay of Fundy. Driven by onshore and southwesterly alongshore winds, these slope water masses occasionally pass from the deeper waters and through the Fundian Channel between Browns and Georges banks. These warm water masses carry welcome injections of nutrients. However, they also have been an irregular feature over time, though scientists have noted a correlation between the occurrence of warm water masses and periods of increased marine productivity.

While most of the Atlantic Ocean is quite barren, the special mix of factors off the coast of Atlantic Canada at the mouth of the Bay of Fundy makes this area as lush as any in the world. The bountifulness of marine life there is dependent upon the energy of the Bay of Fundy's world-famous tides. From spring to fall this tidal energy works in concert with the light and heat of the sun, along with freshening winds, to produce ideal conditions for the flowering of microscopic marine plants called phytoplankton—the primary source of food in the ocean.

Where the push of Fundy tidal currents in harmony with strong spring and autumn winds combine to produce upwellings of water, nutrients from the seafloor rise to the surface. Such favourable conditions then result in the formation of a stratified water column that will trap these seafloor nutrients in the elevated mixed layer near the surface, where they become accessible to free-floating life forms. It is at the water's surface that the energy of the sun has its greatest impact, exciting the pale blue chlorophyll that is carried within the cells of the microscopic plants, or phytoplankton. The chlorophyll, once energized by the sun, acts as a catalyst, enabling the phytoplankton to use the fertile broth of seafloor nutrients bubbling near the surface. Blooms of phytoplankton created under these conditions are a regular feature at the mouth of the Bay of Fundy, and it is the production of such fertile areas that accounts for 99 per cent of all plant life in the sea.

Similar conditions of nutrient upwelling—the result of action created by currents and winds—occur on the Grand Banks off Newfoundland. There the ocean circulation and production of nutrients is heavily influenced by the cold brackish water mass of the southward flowing Labrador Current. The Labrador Current, which originates near the entrance of Davis Strait, is itself the product of an imperfect union of two northern currents. The two influences can later be identified as the current flows southward off the coast of Labrador, when it splits into three portions upon approaching the northern Grand Bank. As a result, an inshore stream flows through the Avalon Channel; a main branch along the eastern edge of the Grand Bank; while the outer branch flows around the Flemish Cap, situated outside Canada's 200-mile limit.

Along with the Labrador Current, the other water mass that determines the availability of nutrients on the Grand Banks is the warm water push from the Gulf Stream. It makes its entrance off the Tail of the Bank from a southwesterly direction and confronts the colder water along an arc toward the northeast. The hostility which follows creates a distinctive oceanic front. Along this interface a trough of current reversal produces a vigorous mixing and churning of waters, resulting in the upwelling that is vital to the lifting of bottom nutrients. Similar to the irregular influx of warm water masses on the inshore grounds off southwestern Nova Scotia, between Brown and Georges banks, warmer continental slope water is occasionally drawn onto the Grand Banks, further complicating the picture.

While the waters off southwestern Nova Scotia and the Grand Banks are better known for their lushness, other less robust pockets of natural fertility mark the coastal waters of Atlantic Canada. Each marine garden is the product of a similar complex interplay of physical forces that create environments rich in the necessary elements for the growth of abundant marine life.

The seasonal production of countless microscopic sea plants is critical to the welfare of fish stocks and, therefore, to commercial fisheries because they forge the first link between the environment (physical forces) and the hierarchical chain of marine life. Feeding on the rich soup of nitrates and other minerals, as only they can, phytoplankton provide a delectable meal for a host of life forms more highly developed. The vital chlorophyll, in addition to providing the fuel for primary food production, is also responsible for the pale blue-and-green colour of the water as its dominance in the nearly transparent bodies of the tiny plants is reflected in sunlight. The quantity of floating phytoplankton produced at any one time can be best tracked by satellites able to view and scan the world's watery surface. Raw data can be used by scientists as an index to estimate the biomass, or total weight, of these primary producers and to indicate the potential strength of commercial stocks feeding further up the chain. Baby cod, for example, are one of many species that feed on phytoplankton as their first food.

There are two rather distinct food chains. One has as its base diatoms, which are microscopic algae having siliceous cell walls. Diatoms flourish under conditions of bottom upwelling followed by stratification of the water column that is generally found at the mouth of the Bay of Fundy, on the Grand Banks, and other fertile grounds along the Atlantic Canadian coast during the spring and fall seasons. Diatoms feed a line generally termed the "traditional food chain," which includes small through medium-sized shrimplike animal forms, which, in turn, feed those species termed "commercially attractive," such as groundfish stocks of cod, haddock, and pollock. The second branch of the food chain has as its foundation phytoplankton, dominated by bacteria and mobile flagellates. This second branch of the chain is generally prevalent during the summer season, when the warmer weather conditions and less frequent wind storms result in a reduction of both upwelling and the mixing of nutrients in the upper portion of the water column. The potential of the second branch is spent in a complex food web—the microbial loop—which is "inefficient" in its nurture of groundfish because those species of this branch that are attractive to groundfish are of a decidedly lower food value.

Ideally, groundfish spend the latter portion of their first two weeks as free-floating eggs in the upper layers of the stratified water column, where the diatoms multiply. During this time, their nutritional needs are met by feeding on the attached yolk sac. If external conditions have been ideal and the physical forces, including tides, currents, and winds, have placed them in a favourable position relative to their food supply, then their chances of surviving the next stage of development will be greatly increased. The larval stage begins as the eggs, a little more than a millimetre in size, hatch into the wriggling form which is ready to feast on the plankton "pastures" at hand. Phytoplankton is the first meal of these hatched larval forms of groundfish, but they soon show a preference for young copepods and other small crustaceans found at the ocean's surface.

Animal-like plankton, called zooplankton, and flea-sized copepods, the most numerous of marine crustaceans, are the next link in the chain favoured by commercially attractive fish stocks. Drifting in huge swarms, these herbivorous grazers feast on the copious amounts of phytoplankton bobbing on the same wave crests. It is from these tiny torpedo-shaped copepods, or redfeed, that the food web widens to include a host of fish, marine mammals, and seabirds. For example, cod are not the only ones fond of young copepods; the 50 tonne right whale strains copepods from the sea water, while the tiny storm petrel swoops down for a feed from the same supply.

Copepods are also a favourite meal of the shrimplike krill. Krill, too, are a planktonic species, passively floating or drifting at the water's surface. Thus, they are positioned and ready to feast on passing copepods. When copepods are in good supply the krill flourish at a rapid rate. As a result, the teeming crustaceans form a large slick, changing the colour of the water to a pinkish hue. This bloom of krill can set off a feeding frenzy of other species further up the food chain. Herring swarm from below to feed on the krill, while seabirds, including the herring gulls, dive for both krill and fish. Krill also attract other animals to the mouth of the Bay of Fundy and the Grand Banks. Finback and humpback whales can eat half a tonne of krill at one feeding, while seabirds, like the razor-billed auk and puffin, are constantly feeding on fish and krill close to their breeding colonies. Thus, there is naturally existing competition between cod and other ocean species for copepods and other food sources. It is essential to the growth of cod that they satisfy their requirements for this vital ration. Given the challenge of variable conditions found in the upper layers of the warmer water column and the competition for food sources, larval and young cod, among other groundfish, have a precarious beginning.

A second vulnerable stage in the development of cod occurs when the fish graduate to the juvenile stage. This transformation from the larval stage occurs when the cod attains a length of about 4 cm (1.6 in.), signalling the need to seek out a new habitat and a new diet. At this point, the nearly transparent fish settles to the ocean bottom to begin its new life. Indications are that the juveniles prefer either a sandy or gravelly bottom but one not too distant from a rocky hiding place, where they can beat a hasty retreat should predators threaten. The protection afforded by this rough bottom is enhanced by the growth of sponges and treelike corals, adding a montage of nooks and crannies in which to hide.

It is perhaps at the juvenile stage that shelter is most critical. These small cod have been fortunate enough to survive the egg and larval stage, yet they continue to face tremendous odds. It has been estimated that approximately 99.9 per cent of the larvae do not find an appropriate place to settle. Many are taken as prey of larger fish; others settle in waters that are either too deep, too cold, or bereft of the necessary sustenance. Given that the availability of a choice habitat is such a critical factor in the survival of young cod, haddock, and pollock, any permanent damage to the rocky ocean bottom will have serious consequences for the survival and propagation of groundfish species.

The diet and the food supply of juvenile cod in particular change with the adoption of a new life on the ocean bottom. The young fish have a ravenous appetite, expressing a preference for krill, amphipods, and other small shrimplike crustaceans, as well as smaller fish and shellfish larvae. It is just off the bottom that the juvenile cod establishes and works hard to hold its position, beating its tail fin frantically, just to maintain its place against the force of the passing flow. It spends its time keeping a watchful eye for a passing meal. Should some attractive morsel present itself, the juvenile cod will spring up, seizing the food, and then allow the current to carry it backwards for some distance before re-establishing its position in the underwater stream.

By the end of their first year, cod approach 25 cm (10 in.) in length and their diet becomes more flexible. Cod off the coast of Newfoundland and Labrador may measure only 15 cm (6 in.) because of colder water and harsher conditions. Cod are somewhat adaptable, feeding on prey swimming the water column as well as attached or slow-moving benthic, or ocean bottom, forms found on the ocean floor. Capelin, herring, sand lance, squid, shrimp, mackerel, silver hake, and turbot are among the favourite foods of adult cod. Although it has no aversion to making a meal on brittle stars, crabs, combjellies, and even the occasional sea anemone. To digest the latter, a cod will swallow a small stone, which will then be used in the fish's stomach to help break down the tentacle-waving hydroids and other creatures attached to the anemone.

Most other groundfish follow the same patterns of early development exhibited by cod. Spawning patterns are also similar. With most groundfish species—including haddock, pollock, American plaice, witch flounder, yellowtail flounder, turbot, halibut, and silver hake—the female spawns on the ocean floor with the eggs fertilized externally by the male. The eggs then gradually float upward to the surface, where they undergo a two-week development before hatching. The resulting larval forms feed on phytoplankton and zooplankton, including copepods, amphipods, krill, decapods, and a variety of shrimplike shellfish.

There are two notable exceptions. Redfish, for one, are viviparous, meaning the female's eggs are fertilized internally. The eggs develop, even after hatching, in the female's body. As many as forty thousand young are housed in the ovaries until the yolk sac of each is exhausted. The young fish then enter the world already hatched and able to fend for themselves. The other exception is winter flounder. Like the majority of groundfish, the female lays her eggs on the bottom, where they are fertilized. With winter flounder, however, the eggs are found in clusters and these remain on the bottom throughout their development.

Juvenile haddock develop somewhat differently from cod—they remain in the upper portion of the water column for two to three months during the summer before settling to the seafloor. They feed on planktonic forms, in effect continuing their larval diet, while living near the surface. Juvenile haddock also serve as prey for a wide range of larger fish that surface to feed. During early fall, haddock descend to the ocean floor, where they begin feeding on attached and slow-moving creatures. Included in their diet are crustaceans such as amphipods, panadalid shrimp, and hermit crabs; echinoderms, such as brittle stars, sea urchins, and sand dollars; and molluscs, such as snails and clams. Adult haddock, like cod, develop a liking for schooling fish, including sand lance, herring, and argentine, though they retain their taste for benthic forms. Younger haddock are most vulnerable to attack by cod, pollock, and silver hake. As the haddock increase in size, their list of mortal foes change to much larger fish, such as halibut.

When the larvae of pollock—the third of the most commercially popular groundfish species—hatch, they are about the same size as cod larvae, measuring 3.5 mm to 4 mm in length. Unlike cod and haddock, pollock do not exhibit a juvenile stage. The larval stage simply continues growing until it reaches a length of 25 mm to 30 mm, when it takes on most adult characteristics. The larvae waste little time, heading for inshore waters soon after they hatch, where they ply the fertile nursery areas along the coast of Nova Scotia and New Brunswick. Here they are called "harbour pollock" and are known to feed on smaller fish and shellfish. As with other groundfish, the larval stage is very vulnerable to predation by larger fish. Adult pollock feed throughout the water column, from the seafloor to the surface. They have a ravenous appetite, feeding on herring, haddock, hake, and juvenile cod, though they often turn their attention to squid, shrimp, and other swimming shellfish which frequent the surface waters.

Redfish are gaining popularity as a commercial groundfish species. Young redfish measure about 7 mm in length at birth and spend the first portion of their life feeding on the prevalent planktonic forms found in the surface waters. They grow rather slowly by comparison with other groundfish species, but upon reaching a length of approximately 25 mm, they move into deeper waters, favouring areas with rocky or muddy bottoms. Redfish feed on a variety of invertebrates and small swimming fish. They, in turn, are prey for cod and turbot.

The flatfish—witch flounder, winter flounder, yellowtail flounder, American plaice, and turbot—follow the groundfish pattern of feeding on planktonic forms during their larval stage. Due to the very small mouth of the adults, the range of food is restricted. Witch, winter, and yellowtail flounder favour a variety of marine worms, though they will also eat small crustaceans and shellfish which they find on the seafloor. American plaice show more variety, feeding upon fish, including capelin and lance, and upon benthic forms, including sand dollars, brittle stars, shrimplike crustaceans, and polychaete worms. Turbot, which frequent the water off southern Newfoundland and Labrador, favour capelin, while those that swim the deeper waters show a preference for shrimp. As turbot grow larger they show a tendency to feed on larger specimens, including cod, redfish, and even other turbot.

Competition for the next meal is very keen among species both up and down the food web. Scientists have made considerable effort to determine what happens to the food chain when one species is fished down or suffers from elevated levels of predation by other species. One would assume that the direct effects of fishing would be to reduce the population of the target species and the by-catch species; this is generally what occurs over the short term. The long-term effects of fishing effort and predation are, however, much less certain. The short-term reduction in population of both the target and by-catch species by commercial fishing effort can be followed by indirect increases in other populations that would be prey for the commercial species, and sometimes even in the target population. The occasional increase in population of a target species at the same time as it is fished harder makes long-range predictions for future stock numbers unreliable. Given financial restraints, the scientific community has yet to develop sufficient mathematical modelling to determine the effect on stocks of changes in patterns of predation and commercial harvesting. It has only recently begun to study the impact of habit degradation by fishing gear upon fish populations.

At best, the analysis of commercial catch and scientific survey inputs can provide only an approximation of the number of fish swimming the ocean and does little to predict the effects of perturbations. Those studying food webs and the ecology of marine communities acknowledge that it is nearly impossible to predict the consequences of fishing down a particular commercial stock and the related impact on its predator, prey, and other competitor populations. However, the only time that predictions of future stock health can be made with any degree of certainty is when a stock is fished down to a point where further effort would unquestionably produce a steady decline in the adult spawning population. This, of course, would result in a decrease in reproduction and a steady decline in the number of new fish recruited to the stock. Fished hard enough, a target species can be reduced sufficiently to cause a population collapse, meaning a plunge to levels where it is no longer feasible to fish commercially (such is the case with northern cod). Eventually, annihilation of the species is the result.

Despite the best efforts of some researchers to study the effects of changes in marine food chains and effects on marine life, there remains a lack of basic knowledge about the interrelationships of the food web in the Canadian Atlantic. Funded scientific work has been channelled into other priorities, such as measurement of stock size and quota assessment, while fundamental research has suffered. For example, more is known now about the food preferences of commercial species, but little is known about the intricate underpinnings of the food web which, in turn, support these predator-prey relationships. A greater understanding of these relationships would permit more accurate predictions for stock health and population.

Posing an equally difficult challenge for the scientific community is the attempt to discover critical factors in the secretive reproductive life of valued groundfish stocks, such as cod and haddock. Concern has been expressed for more than a century that the trawling of fishing grounds during the spawning season may be detrimental to the reproduction of groundfish. These fears have not dissipated over time, yet little more is known today than when the issue was first raised. The deeps do not offer up their mysteries casually. The facts of cod procreation—that most commercially popular species—are still unavailable, except for some important studies carried out on fish in captivity. It appears that cod are bashful creatures and consider their mating a very private matter. Given the choice, codfish show no inclination to perform for scientific observation.

What is known is that the average female cod has vast reproductive potential, capable of laying over a million eggs per year. Each subsequent year she will lay even more eggs. Should she escape the fishing net and other predators and reach old age, she will carry a maximum of 9 million eggs in her final year. A biologist once calculated that if all the eggs spawned by all the ripe female cod in one season survived to become adults, then the entire ocean would become a mass of twisting and turning tails and fins. However, even without the intense fishing effort across the world's oceans, few cod survive a harsh environment and the voracious appetite of all those other species that eat them.

Generally, the species that produce the largest number of eggs, such as cod and haddock, suffer the greatest losses. It seems nature's way, to produce greater numbers of eggs to compensate for the difficult road ahead. If only two eggs of each female's clutch were to survive to adulthood, the stock would maintain its numbers; a lesser success rate would send the stock into decline. So, this remarkable reproductive potential is easily threatened, and any untoward pressure by humans or unfavourable environmental developments could jeopardize the survival of a species.

Studies relating to the spawning of cod date back to 1876, when Professor Sars of Denmark published a paper titled, "On the Spawning and Development of the Codfish." His follow-up study three years later indicated that after spawning, eggs from the codfish floated to the surface of his aquarium. This result he extrapolated to the natural environment, suggesting that the eggs would rise above any possible disturbance of trawls working the ocean bottom. A number of papers and books published subsequent to Sars's two papers credit his work with diffusing the growing voice of suspicion and fear from fishermen that trawling was causing an abortion of the natural spawning process.

However, the definitive study on cod reproduction behaviour thus far remains a 1961 report by Vivian Brawn carried out at the University of Durham, England. Her twenty-two-page report, published in the prestigious scientific journal *Behaviour,* describes the courtship ritual performed by male and female cod in captivity. The dominant male marks off his territory and awaits the arrival of a female into his domain. Unreceptive females and other males that may present a challenge are driven away.

The female initiates the courtship response in the male, which responds in kind with a flaunting display. Body language is key to the response as both male and female follow a prescribed ritualistic "dance," although some communication made by the male involved a low grunting sound. This sound was also noted in a later study in 1976 by Hawkins et al., who recorded the sounds made by male haddock involved in the courtship of females. With both cod and haddock, the male follows the receptive female around in a defined space within the aquarium, resulting in sexual contact. The eggs and milt are then released as both swim upwards, belly to belly.

A pertinent aside from the Brawn and the Hawkins et al. studies is the discovery that during the pre-spawning process, some of the cod were "extremely frightened" by the falling food provided by the handlers. "A male and female on separate occasions fled in such panic that they collided at full speed with the walls of the tank." This indicated that the fish were much more nervous during spawning than they were during the rest of the year.

This mating ceremony observed in the laboratory has been witnessed in the ocean and on a grand scale. George Rose, a scientist with the Department of Fisheries and Oceans (DFO) in St. John's, Newfoundland, reported in 1993 that a research vessel had, the year before, tracked one of the few remaining huge schools of northern cod. The spawning cod were observed to have come together near the seafloor and to have moved along what was thought to be an established migration route off the east coast of Newfoundland. The school, which consisted of hundreds of millions of cod jammed fin to fin, were discovered at a depth of about 305 m (1,000 ft.), situated just north of the Grand Banks. The school measured roughly 2 km (1.5 mi.) across and was shaped like an inverted saucer. The aggregation of fish remained rather inactive over the first ten days of observation, hovering near the seafloor, though occasionally a midwater spawning column would be detected on the sonar. This was thought to represent a number of pairs of male and female cod that had initiated courtship near the bottom and were then involved in the ritualistic sexual union, taking them aloft to mate above the press of bodies below.

In Pol Chantraine's book *The Last Cod-Fish* reference is made to a Norwegian study on the reproduction of cod retained in a large netted cage located in the sea. Night-vision television cameras were used to record the cods' interaction. The fish behaviourists noted that the male cod builds a nest for his mate. He then busies himself with keeping the area spotless, swallowing any foreign objects that might enter a defined water column that he has staked off. Having cleaned the nest area, he awaits the female. Should even the smallest piece of foreign matter drop into the designated area, the male is terrified and immediately swims away to begin the process in another location. The study shows how easily the male is put off, how fragile the right conditions for reproduction are. The detrimental impact of invasive fishing methods becomes clearer in light of this fragility.

Once the mating has occurred, the survival of the eggs becomes a concern. The eggs of most groundfish are pelagic, that is, they eventually float to the surface or uppermost layers of the water column. Not all key commercial species produce pelagic eggs and some eggs, such as those of herring and winter flounder, are known to settle to the bottom after spawning, where they are found in clusters on the seafloor. Other species whose eggs are found on the sea bed include sand lances, skates, and rays. Eggs that lay on or float near the seafloor before hatching face a hostile environment. They run the risk of being buried and smothered by material moved by the natural forces that shift the ocean floor's surface.

Furthermore, fish eggs are oxygen dependent throughout their development; therefore, the dissolved oxygen level in sea water in which the eggs bob is important. Should sand and silt clouds be created by disturbances along the bottom, the resulting clouds deplete the oxygen supply and endanger the low-lying eggs. Fishing methods which employ powerful groundfish draggers, scallop draggers, or clam dredges further exacerbate an already tenuous situation for these life forms as they rip and tear at the seafloor, rolling rocks and boulders, and creating sand and mud clouds.

The eggs of cod, haddock, pollock, witch flounder, yellowtail flounder, and American plaice are considered pelagic. It has been assumed, therefore, that once they float to the upper layers of the water column, they remain there, although they are then at the mercy of the tides, currents, and winds. However, DFO studies conducted in 1990 of egg and larval stages of haddock and cod found off southwestern Nova Scotia have overturned the assumption that pelagic eggs spend their entire pre-hatch phase on the surface. While the early egg stages have been shown to gradually float upward through the water column, the same pattern does not hold for the stage prior to hatching. Eggs in the last stage—twelve to nineteen days before hatching—are found in the greatest numbers at up to 50 m (164 ft.) beneath the surface. Although the point of greatest density of egg mass varies according to the growth stage and the salinity of the egg, something as variable as heavy rainfall would make the eggs less buoyant, causing them to sink to depths approaching the ocean floor. Eggs are capable of refloating to the surface, but the studies showed that rough seas drive them under again.

The same series of DFO studies showed scientists that groundfish prefer a specific kind of "hard bottom" with sand and gravel (called Sambro sand), and pebbles "on which to lay the eggs." Hard bottom is a term typically used to describe that ground characterized by boulders, ledges, and other rock outcroppings. The fact that fish lay their eggs on the bottom, showing preferences for particular types of sand and pebbles, is seldom acknowledged by scientists, yet has been long acknowledged by fishermen. Many scientists have maintained that the female's eggs are fertilized as they are expelled while the male and female mate at some distance above the ocean floor. The suggestion follows that these eggs never find themselves on the seafloor unless, of course, they have first expired, having died in a harsh environment—a change of temperature, too much or too little salinity, a lack of dissolved oxygen, or some other benign stroke of nature. If, in fact, the eggs of cod and haddock are laid in the sands and pebbles of the bottom then, of course, they are fatally susceptible to burial by fishing gears and the appetites of hungry filter-feeding bottom-dwellers.

Upon reaching adulthood, groundfish become active migrants capable of seeking out preferred environments, following the seasons and actively pursuing ideal water temperatures and the bloom of prey upon which they are dependent. In summer, cod frequent inshore waters to the edge of the continental shelf along the Atlantic coast—from Greenland in the north to waters off Cape Hatteras in the south. Like other groundfish, cod are partial to ideal water temperatures between 0°C and 4°C. In the fall they return to deeper waters as the inshore temperatures drop. It is in these deeper waters that they eventually mate and spawn.

While our knowledge of groundfish habits is rather limited, much of the impetus for studying the migration and distribution of fish stocks has come from harvesting interests concerned primarily with maximizing their take. Government research initiatives have, in the past, included support for the harvesting interests as every effort was made to develop a competitive Canadian fishery. Basic research necessary to determine suitable levels of harvest, which should have been carried out prior to the commencement or expansion of a fishery, often took a back seat to the corporate agenda. Under pressure from the private sector, controls and research which should have been carried out to help ensure sustainable levels of harvest that were often abandoned.

Research into the habits of groundfish indicate that in addition to depth of water, salinity, type of bottom and bottom sediment, and even latitude, a preferred range of water temperature also influences the distribution of stocks. Cod show a distinct preference for water between 0°C and 4°C. Fish taken from the optimum range and placed in warmer or colder water quickly respond by swimming back to a depth exhibiting the ideal temperatures. Fishermen use this knowledge in selecting optimum water depths in which to fish and set their gear.

The temperatures of the ocean vary with water depth. The first 10 fathoms of inshore water is warmed by the summer sun to a balmy 5°C to 15°C—too warm for adult cod. The ideal range in temperature is found at depths of 10 fathoms to 40 fathoms, and it is here that cod traps and handlines are most effective. In depths of 40 fathoms to 100 fathoms, the water temperature drops to -1°C—too cold for cod and other groundfish. Below 100 fathoms the water temperature returns to the ideal 0°C to 4°C. The interface between the cold intermediate water and the warmer deeper water is best suited for gillnets and longline fishing. Since the warmer temperatures of the deep water remain nearly constant year-round, offshore trawlers can fish through the dead of winter. The gear is lowered through the cold upper water layers to fish the warmer water some 200 fathoms below the surface.

Cod preparing to spawn, as they generally do from early to midspring, avoid water cooler than 0°C and demonstrate preference for temperatures between 2.5°C and 4°C. In the days before sophisticated fish finders, information on favoured water temperatures was key to increasing the catch and reducing the effort required as larger seaworthy vessels using mobile gear targeted spawners amassed in deep, warmer winter waters. It is when the cod and other groundfish are gathered in a defined and accessible area that the trawler is most efficient in its catch. The growing plea from other gear sectors, such as the inshore fixed and mobile fleets, is that the fishing of such staging and spawning areas—where fish are gathered, dormant, in groups— must be controlled or eliminated to conserve the stocks and provide more reasonable access by other harvesting interests.

Migration patterns determined by these environmental factors are important to fishermen in anticipating the location of fish. There are twelve stocks of cod along Canada's Atlantic coast. Each is an identifiable unit exhibiting unique characteristics, though there is some intermingling between the adjacent stocks. The cod of some of these stocks are known for their lengthy migrations.

In the eastern Gulf of St. Lawrence concentrations of pre-spawning cod found off southwest Newfoundland migrate northward in February and March, largely under ice-covered waters to a position along the west coast of Newfoundland and the north shore of Quebec. In the western Gulf of St. Lawrence, concentrations of cod move back and forth between winter areas

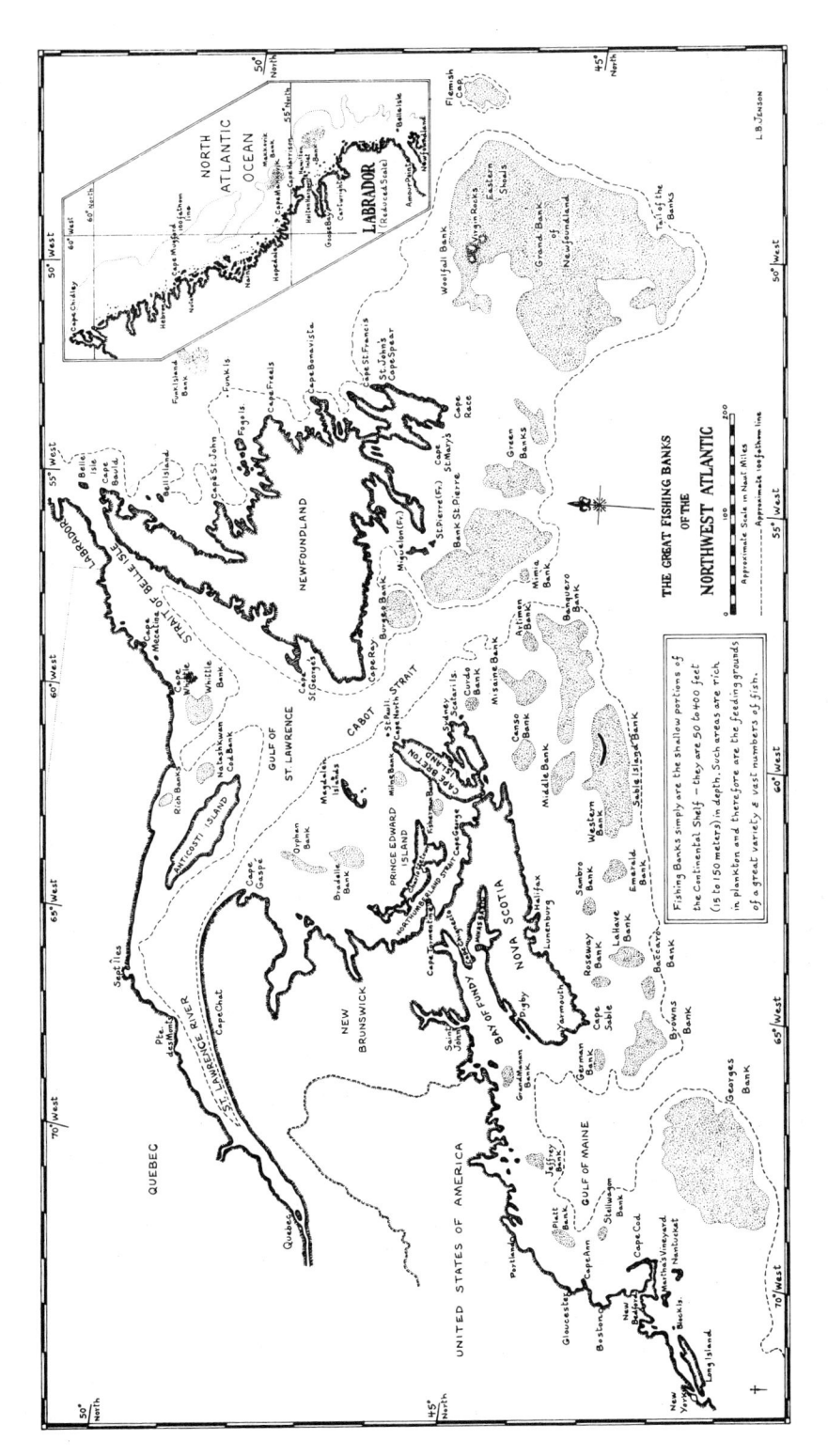

The fishing banks of the Northwest Atlantic on which the Atlantic groundfishery was so dependent.

northeast of Cape Breton Island and summer areas off the Gaspé and the Bay of Chaleur.

Spawning areas are not as distinctly marked in waters off Labrador and Newfoundland as they are off southern Nova Scotia and New Brunswick, where Browns and Georges banks are located. Spawning areas off Labrador and Newfoundland have been identified off the northern Labrador Shelf, off Cape Chidley, on Hamilton Bank, Bell Isle Bank, Funk Island Bank, Flemish Cap (outside the 200-mile limit), the eastern and northern slopes of Bank St. Pierre, the west coast of Newfoundland, and along the Quebec north shore. It is to these spawning areas that cod migrate in winter and early spring to reproduce.

Northern Cod off Labrador and Newfoundland are thought to consist of three races or stocks of the species, each with its preferred spawning and migratory patterns. The Labrador stock spawns on Hamilton Bank, later migrating to feed along the coasts of Labrador and northern Newfoundland. The second stock, termed the northeast Newfoundland stock, also spawns on Hamilton Bank and additionally southward to Funk Island Bank. In summer this stock feeds along the coast of northeast Newfoundland. A third stock spawns on the northern Grand Bank and feeds along the coast of eastern Newfoundland in summer. The eggs and larvae of these stocks are known to drift in a southerly direction on the Labrador Current where, under ideal conditions, they are surrounded by a ready food supply.

A number of cod migration patterns also exist for stocks off Nova Scotia. Cod from Banquereau, Miscaine, and Sable Island banks off the eastern part of the province are known to migrate to the inshore areas with some stocks actually entering the Gulf of St. Lawrence. Stocks off southern Nova Scotia tend to spend the winter either in deeper inshore waters or on La Have and Browns banks. In both cases, the cod seek out more shoal waters in the summer with inshore fish moving to adjacent shallow waters, and fish on the banks moving to the shallow areas of the offshore banks. Georges Bank cod remain mostly at home and congregate in shallower waters on the bank in summer.

Haddock, a prized species for the fresh-fish trade, has been under tremendous pressure for years. Largely fished out in waters off Newfoundland, this species is found predominantly in the Gulf of Maine, Georges Bank, and Browns Bank areas off New Brunswick and Nova Scotia. Haddock are typically found in water depths of 5 fathoms to 110 fathoms, from inshore waters to the edge of the continental shelf. In summer, concentrations occur in the Bay of Fundy, along the southwestern coast of Nova Scotia and to the coast of Cape Breton Island. During the winter there is an offshore migration to Georges, Browns, and Sable Island banks. It is on Georges, Browns, Emerald, Grand, and Scotian Shelf banks that spawning activity occurs, beginning in late January and continuing to July. The normal pattern of migration calls for ready individuals to amass in staging areas near the offshore banks early in the year. It is there that they await a warming of the water before moving onto the banks to spawn.

Pollock has not received nearly the amount of attention or study afforded cod, the "king" of the groundfish. Much more work needs to be done to more adequately map the migration route of this species. Research has shown that pollock off Canada's Atlantic coast spawn in late fall and early winter. Mating usually begins in late October, triggered by a cooling of water temperatures to 8°C to 10°C. The peak of spawning activity occurs from November to February, when water temperatures have fallen further to 5°C to 6°C. Scientists have discovered that important pollock spawning areas are found in several locations along the Scotian Shelf, Browns Bank, the northeastern peak of Georges Bank, and Jefferys Ledge in the Gulf of Maine, with ready adults gathering in these areas in the fall of each year. Juvenile pollock busy themselves by frequenting the harbours and bays along the New Brunswick and Nova Scotian coast before returning to deeper waters to live out their adult life.

Redfish range throughout the continental shelf waters of Atlantic Canada. They prefer the cooler northern waters off Newfoundland and Labrador but are found throughout the region from the Gulf of Maine to waters off Baffin Island in the north. The ideal temperature range is from 3°C to 8°C, and the ideal water depth ranges from 15 fathoms to 125 fathoms. Smaller redfish generally seek out shallower water, while larger members prefer deeper waters. Redfish are known to feed in the upper portion of the water column at night, settling back to the bottom by day. This behaviour is taken advantage of by fishermen who generally concentrate their bottom trawling efforts during the daylight hours, and employ large midwater trawls during the nighttime, enabling the fleet to engage in a twenty-four-hour fishing effort. Redfish frequenting the Gulf of Maine and Scotian Shelf waters are not thought to be active migrants. The stock found in the Gulf of St. Lawrence is, however, thought to undergo a winter migration to the Cabot Strait with a return to the Gulf in the spring.

The flatfish generally thrive in the cold, deep waters of the Canadian Atlantic, although there is considerable variation between the species. These fish are taken by trawlers specially equipped with heavy bottom gear designed to scare the flatfish from their resting place along the bottom and upward into the path of the passing net. For example, the witch flounder is found in the deep holes and channels situated between the coastal banks and along the edges of these plateaus, where a suitable range of temperatures is available. That ideal range is from 2°C to 6°C. Preferred water depths range from 30 fathoms to 65 fathoms. The witch flounder is not known to make long migrations and appears to move only when it is necessary to maintain its favourite water temperature. Pre-spawning concentrations of witch flounder occur in the late winter and early spring. Spawning occurs along the eastern coast of Newfoundland, in the Gulf of St. Lawrence northeast of Prince Edward Island, and along the Scotian Shelf to the Gulf of Maine.

The American plaice is no doubt the most abundant flatfish in the northwest Atlantic and ranges throughout the Canadian territory from the Gulf of Maine to Baffin Island. It prefers cold water but frequents a range from -1.5°C to 5°C or more. This species is found from shallow coastal waters to depths in excess of 100 fathoms. Its area of greatest concentration is the southern Grand Banks off Newfoundland, though other established grounds include the waters off southern Labrador, the edge of the Laurential Channel in the Gulf of St. Lawrence, and along the outer edge of the Scotian Shelf off Nova Scotia. No specific spawning grounds have been identified for the American plaice.

Yellowtail flounder does not seek out the deeper waters, as witch flounder and American plaice do, preferring depths of less than 50 fathoms. Its ideal temperature range is from 3°C to 5°C. The Canadian range of this species is from the Strait of Belle Isle off the northern tip of Newfoundland to the Gulf of Maine. The greatest concentration is, however, on the Grand Banks off southern Newfoundland where most of the spawning for the east coast stock occurs.

Winter flounder, too, prefer the more shallow waters along the Atlantic coast and on the banks of the continental shelf. This species is more temperature tolerant, with stocks in the southern part of the range along the Scotian Shelf showing a preference for temperatures ranging from 11°C to 15°C; stocks off the coast of Newfoundland show a preference for temperatures below 11°C. This is one species which can tolerate waters of low salinity, enabling it to stake off a niche free of competition from other flatfish.

Turbot, or Greenland halibut, favour the cold northern waters off the east coast of Newfoundland and Labrador and are found in the deep waters of the continental slope. Deep-sea trawlers fish turbot from the Nose and Tail of the Grand Banks to the deep-water channels off west Greenland.

Since groundfish are typically found feeding on and occupying that portion of the water column near the bottom or "ground," they expose themselves to modern fishing practices that are increasingly capable of fishing stocks to the point of collapse. Basic knowledge about preferred temperature and depth ranges of groundfish, coupled with information on the ocean floor terrain, provide fishermen with critical information, useful when searching for concentrations of different species. Fishing gears are designed and manufactured to exploit established fish habitats, rendering the fish easier and easier targets as technology advances. For example, large cod, the prize of fixed gear fishermen who are the best equipped to fish them, are generally found on rough bottom. The shallow-water banks with smooth bottom pose little or no threat to the towed gear of trawlers and draggers. These grounds usually feature a higher proportion of smaller cod, an assumption borne out by observation of the landed catch. Similarly, the flatfish show a preference for flat bottom and thus are susceptible to the specially designed gear of draggers which root them out of their partially buried condition as they rest and feed on the bottom. Meanwhile, redfish, which frequent deeper waters, are rather easy targets for draggers as they swim about in large schools. They are taken during the day by bottom trawls and at night by midwater trawls.

Catch data shows that despite persistent pressure, fish are creatures of habit. Even with their numbers greatly reduced by fishing, they continue to congregate at their favoured spawning and nursing grounds, intuitively drawn to those comfortable environs they are seeking. The continued predictability of their habits leaves stocks vulnerable to persistent heavy fishing pressure once they are discovered.

It is said that a little knowledge is dangerous. This is certainly the case with respect to our ocean environment. Our fishing policies and quotas are based on inadequate information about the vagaries of such a complex and intricate web. The promise or failure of a sustainable groundfishery hinges on our understanding of the important roles of physical and biological forces at work in the underwater world that we exploit. On the other hand, the facts that have been gleaned by our scientific community have proven lethal in the hands of our fishing industry that is driven to maximize its take of ocean resources. To better understand and respect the laws that govern marine life is to possibly hold the key to renewal and sustainable development of a potentially vast resource. To this end, responsible effort must be adopted by all sectors of the industry, including inshore and offshore fleets, fish plants, and offshore corporations.

Chapter Three

Blaming Nature: In Search of Environmental Scapegoats

By 1984 inshore northern cod catches off Newfoundland had dropped significantly, and fishermen were becoming more and more vocal about declining stocks. Inshoremen were among the first to challenge the government perception that northern cod stocks were healthy. For over five hundred years, the Atlantic Ocean had swarmed with fish; the shift away from that abundance had become too drastic to ignore. Fisheries managers were under fire to account for the decline and to devise strategies to stop it from going further. Rather than accept sole responsibility for mismanagement of the resource, government focused on research into environmental circumstances that had contributed to the decline. The Canadian fisheries management system as developed in the late 1970s and early 1980s was reputed to be the best in the world. With such a high-profile reputation it could not readily concede to fumbling its mandate as steward of the ocean's resources. The research conducted from about 1988 to 1993 does provide evidence to suggest that environmental factors and imbalances in predator-prey relationships have played a role in the decline of fish stocks.

Although various aspects of the marine environment can be singled out for study, the results of interaction between physical and biological forces often defy explanation. The ocean environment is very complex and highly changeable. Correlations between environmental factors and the impact on marine life are possible, but lack of a definitive cause-and-effect relationship increases the probability of flawed predictions with relation to their impact on these complex marine systems. Nevertheless, such predictions have been used in determining groundfish stock quotas and stock health for the Atlantic groundfishery.

Historically, scientists have been left with the impossible task of explaining what is often unexplainable. By 1988 Newfoundland's cod trap fishermen expressed such concern over the decline of inshore catches that scientists and fisheries managers came up with a theory hoping to allay their concern. The explanation was a well-meaning attempt by federal officials to calm growing fears expressed by inshore fishermen that a total collapse of the northern cod stock was impending.

Backed by the findings of DFO research cruises and reports that the offshore trawlers were finding a ready supply of northern cod, federal scientists were convinced that the stocks were still growing in 1988. They did acknowledge the decline in the inshore catch, but they looked to environmental changes to explain the troubling signs. They did not entertain the possibility that failed inshore catches were the direct result of increased harvesting by the offshore fleet, which was the argument of inshoremen.

One suggestion was that northern cod migration patterns had been affected by a wider than normal barrier of very cold water off the coast of Newfoundland, which discouraged the inshore migration of cod. It was reasoned that the cod did not have the necessary resolve to swim the extra distance through the unusually cold waters to reach the harbours and bays of the Newfoundland coast as they had done for years. Instead, the northern cod appeared to be expanding its traditional range and travelling further south.

The second possible explanation offered by the DFO was any cod that did make it to the inshore zone found the waters near the cod traps too cold to venture near. A change in the water temperature gradient—the result of local winds and currents moving the warmer water off and the colder water onto the inshore grounds—discouraged northern cod from frequenting the harbours and bays where the inshore trap gear was set. Cod show a preference for water between 0°C and 4°C, avoiding water much warmer or colder than this.

A third suggestion to account for falling inshore northern cod catches was possibly reduced cod fishing effort by the inshore sector. In 1986, for example, there were three times as many capelin licences issued along some parts of the Newfoundland coast as there were in 1985. It was believed that inshore fishermen were turning their fishing effort to capelin or crab and away from cod.

Scientists studying the collapse of northern cod in the late 1980s have exercised considerable faith in the connection between the unusually cold water temperatures and the decline of the stocks on Atlantic Canada's coast. While average temperatures have been increasing slightly over the North American continent and over much of the northwest Atlantic, they have shown a decline in waters from Greenland south to northern Nova Scotia. For about thirty years the northern-most range of northern cod has suffered from less than ideal cold conditions. More recently this chilling trend has made its way south to challenge the northern cod in its traditional waters off Labrador and Newfoundland.

This cooling trend is explained by Canover et al. in a 1994 report from the *Proceedings of Coastal Zone Canada* as a consequence of global warming and an increase in cold winter winds blowing over waters off the shore of western Europe. These northwesterly winds disturb the normal subarctic flows in the northeast Atlantic, forcing inordinate amounts of cold polar water south, where they are picked up by currents east of Greenland and Iceland. As well as producing a chilling effect, the movements bring more ice floes into the melt zone, thus decreasing the salinity of the ambient sea water. The environment's ideal conditions are further disrupted because this low salinity cold water is less likely to mix with deeper warmer waters. This abnormal stratification prevents the usual upwelling action that carries nutrients critical to the bloom of phytoplankton to the water's surface.

The existence of this cold, low salinity water mass and its effect on nearby water and climate were first noted in 1961, when it was named the "great salinity anomaly." Its initial path was tracked on oceanographic charts as it started its counter-clockwise, subarctic, circular movement through waters north of Iceland. It was not until 1973 that its impact was felt on the Grand Banks off Newfoundland. At that time, a number of groundfish stocks suffered poor year classes, meaning fish did not survive, because the production of phytoplankton and zooplankton fell through the early 1970s.

It is believed that a similar saline anomaly is also partly responsible for the plunge of cod numbers on the Labrador Shelf and Grand Banks. Reports indicate that both water temperature and salinity are unusually low and perhaps worse there because of intensified global warming. Increased melt from glaciers and freshwater runoff from rivers in the north are also perceived to be the results of global warming. The resulting flows add to the vast amounts of cold, low salinity waters already infiltrating the Labrador Current from the Canadian High Arctic, the Hudson Strait, and the East Greenland Current. This cold, low salinity water mass appears to dominate the Labrador Current, making it less likely to mix with the warmer Atlantic waters underneath. The resulting stratification of the water column hinders the mixing of nutrients from deeper water in this area as well. Northern cod used to gather each spring in the deeper warmer waters of the Hamilton Bank off the coast of Labrador to mate and spawn. Based on the movement to warmer water, fisheries managers speculated the northern cod stock off Newfoundland was migrating southward in search of ideal conditions.

With the presence of the salinity anomaly, the eggs produced by northern cod hatch in colder layers with fewer nutrients. The young cod are hungry for copepods. There are three species found in the pelagic zone: *Calanus finmarchicus, Calanus glacialis,* and *Calanus hyuperboreus.* The favourite is *Calanus finmarchicus.* However, the colder, low salinity water promotes the growth of the less desirable copepods—*Calanus glacialis* and *Calanus hyuperboreus.* The result is that the preferred food of young northern cod may be overshadowed by its cousins when the pelagic zone experiences below normal salinity and temperatures. Populations of the favoured copepods may also be lower because *C. finmarchicus* is even more sensitive to cold than cod larvae, so they may be significantly delayed in their hatch and growth in the colder water. Mismatched cycles occur, with the favoured copepods hatching too late to suit the needs of larval cod.

The mismatched cycle in the food chain may influence groundfish stocks frequenting the Hamilton Bank and waters off the east coast of Newfoundland. Cod spawned on the more southerly Grand Bank and Bank St. Pierre have traditionally hatched as much as two months later and have not been affected by colder water or lower salinity. The eggs hatched on the Grand Bank and Bank St. Pierre under more favourable conditions, make their way to comparatively warmer surface waters, where the chances of finding the preferred copepods are much greater. It may be assumed that cod hatched on the southern Grand Bank and Bank St. Pierre would find an adequate food supply and that stocks native to the area would be more successful. In fact, groundfish off the southern coast of Newfoundland have shown marginally more resilience than those swimming the colder, more northerly waters. The more southerly fishery of the Grand Banks and Bank St. Pierre remained open until September 1993, a full year longer than the main northern cod stock fished from ports in Labrador and northwestern Newfoundland.

Older, more mature cod extend the spawning season for the stock because they spawn later in the season and over a greater period of time than younger cod. However, there has been a dreadful drop since the late 1980s in the proportion of more mature cod to younger cod as fishing effort and capacity increased. As the more fecund female fish become increasingly harder to find, the spawning season for all stocks grows shorter. Hence, the chance increases that the production and hatch of cod larvae will not match the production of their preferred copepods. The once synchronized cycles have been effectively thrown off.

Colder waters also may have affected the food supply of adult northern cod. There are indications that capelin are now ranging further south in search of nourishment. They, too, are dependent upon copepods and other small crustaceans that now occupy more southerly waters. The adult cod are occupying water outside their traditional range because their food source has moved, supporting the theory of a southern migration of the stocks.

While the relationship between changing environmental conditions and the decline of the northern cod stock off the coast of Labrador and Newfoundland has gained most of the public attention, other groundfish stocks appear to be in a very critical situation, possibly even collapsed. Scientists have found that the remaining groundfish in the waters off Newfoundland and Labrador, including American plaice and witch flounder, are following the same trend as the northern cod. They inhabit a much narrower range than before and show a preference for deeper waters—signs that indicate stocks are under severe stress. Near record cold air temperatures have blasted the Labrador-Newfoundland region for the past ten years, and this appears to have further reduced commercial and non-commercial stocks. This cold air, combined with strong northwesterly winds, has resulted in early ice formation. The extended ice coverage in waters off southern Labrador and northern and southern Newfoundland, makes surface waters colder. In fact, in 1992 new records were set for the duration of ice coverage in areas on the outer reaches of the northern Newfoundland Shelf.

Similar weather conditions have persisted for Gulf of St. Lawrence waters. Cold winter temperatures have combined with strong northwesterly winds to lock the region in near record ice coverage for the past ten years. Cod stocks in the Gulf remain at the lowest level ever recorded. Also in decline are redfish stocks which frequent deeper waters; white hake typically found on the inshore grounds; and American plaice found in the southern portion of the Gulf. The colder waters have also forced groundfish to adjust their living patterns. The Gulf of St. Lawrence cod stock has, over the past ten years, been found in waters further and further south and may now be at the most southerly limit of its range in the gulf.

The groundfish stocks in waters of the Scotian Shelf and Gulf of Maine have suffered somewhat from the influence of below normal winter air temperatures through the late 1980s and early 1990s. Heavier than normal ice conditions have been experienced as far south as Halifax, and many bays and inlets along the Nova Scotia coast have been locked in ice for longer periods. While coastal Nova Scotia waters have been colder, the deep waters of the basins and Cabot Strait off Cape Breton have experienced warmer than average temperatures. These warmer and deeper waters found in the basins and channels of the continental shelf are believed to have benefited from the influx of warmer offshore slope waters. Although surface waters along the Scotian Shelf and in the Gulf of Maine have been colder, they have not approached record lows. Also, the temperatures of the water at depths frequented by groundfish have not dropped below the ideal range. In fact, a slight reduction of the ambient water temperature seems to have improved conditions for those commercial species near the southernmost limits of their range. Gulf of Maine plankton production has been up, including zooplankton (copepods and other small crustaceans), improving the feeding conditions for larval stages of groundfish.

Scientists and managers responsible for the Scotia-Fundy region situated off southern New Brunswick and the southern and eastern shore of Nova Scotia published a 1994 study titled Report of the Workshop on Scotia-Fundy Groundfish Management from 1977 to 1993. Angel et al. concluded that although there has been a reduction in the biomass of groundfish in the region and the size and age range of the spawning population is dangerously low, these conditions are less a result of environmental conditions and more a result of increased fishing pressure.

The same is not the case, however, for stocks in waters off Cape Breton and the eastern portion of the Scotian Shelf. In these waters a combination of harsh environmental conditions and excessive fishing effort has resulted in a decline in the populations of all groundfish including cod, haddock, pollock, and flatfish. The decline in water temperatures in the Scotia-Fundy region has also been accompanied by a change in migration routes of groundfish and their prey, such as herring and mackerel.

Just as planktonic food like copepods are essential to larval and juvenile cod, adult cod are affected by the availability of their preferred prey species, such as capelin. Northern cod and Atlantic cod, found off the southern coast of Newfoundland, express a definite preference for capelin. Considerable alarm has been expressed lately over the parallel decline in offshore capelin stocks and these cod stocks. The decline may be a result of both cod and capelin responding to the same physical impacts, but scientists are not sure of this.

At the same time cod and capelin stocks are in decline, seal populations are increasing. Fishermen familiar with the eating habits of seals know their voracious appetites for cod and capelin, and adult cod and seals are competitors for the shared food supply. Hence, fishermen have drawn attention to the probable correlation between the explosion in the seal population and the continued decline in groundfish stocks.

The harp seal population off Newfoundland has shown steady growth since 1983. The harp seal is a formidable competitor with cod in the hunt for capelin. Although capelin is the primary prey of harp seals, reports indicate that since 1990 cod has become the most important prey species of near-shore harp seals in Newfoundland. Fishermen are convinced that the growing number of seals is a significant factor in the continued decline of groundfish stocks, and they therefore support the need to re-establish a seal cull.

The harp seal harvest was terminated in 1983, when the European Community responded to the anti-sealing lobby by imposing a ban on the importation of seal products. Fishermen argue that with a moratorium on groundfish catches, the seal herd is free to wreak havoc on remaining stocks. They believe these seals have taken thousands of tonnes of fish that would otherwise have gone to fill their boats and keep the groundfishery going. The continued ban on sealing is thus defeating the purposes of the fishing moratorium on groundfish. The urgent plea of fishermen is that these predators must also be managed.

The most recent estimate, according to a July 15, 1995 article titled "Seals Eat More Cod than Fishermen Catch," in a South Shore newspaper titled *Sou'Wester,* fixes the harp seal population off Newfoundland and Labrador at approximately 4.8 million animals. That is twice the size of the herd estimated by scientists during the mid-1970s and a full 1.8 million more than in 1990, only five years ago. The report also indicated that harp seals are now devouring 6.9 million tonnes of fish and other prey each year. That is double what they took in 1981. The harp seal population is believed to have consumed 40 per cent of that annual total (or 2.8 million tonnes) in waters off the coast of Newfoundland, particularly the northeast coast; 14 per cent (or approximately 1 million tonnes) in the Gulf of St. Lawrence; and the remaining 46 per cent (or 3.2 million tonnes) in Arctic waters. The average harp seal is thought to eat 1.4 tonnes of fish each year, with the majority of the cod ranging from one to two years in age and measuring 10 cm to 20 cm (4 in. to 8 in.) in length. The amount of fish taken would have kept a number of large processing plants open in the early to mid-1980s.

A parallel situation has surfaced with an explosion in the population of the grey seal, which inhabits offshore and coastal waters from the Bay of Fundy to southern Labrador. The greatest concentrations of grey seals are found on the Scotian Shelf, along the eastern shore of Nova Scotia, and throughout the waters of the Gulf of St. Lawrence. Two breeding colonies of grey seals have been identified in Atlantic Canada. A larger colony breeds on the coast and nearby ice floes of the southern Gulf of St. Lawrence during January and February. The other colony breeds on Sable Island, approximately 80 km (50 mi.) off the Nova Scotia coast. Scientists studying the Sable Island colony have found that the production of pups has increased more than 12 per cent, annually, since the study began in the early 1960s. At that rate, the herd on Sable Island is doubling every six years. A 1994 estimate put the Sable Island colony at 82,000 seals, with another 62,000 in the Gulf of St. Lawrence.

Not only are the populations of grey seal herds growing, but their diet coincides with much of the diet of the threatened cod stocks. Related studies have shown that 90 per cent of the food intake of grey seals on Sable Island consists of cod, silver hake, squid, and sand lance. These other species are also on cods' preferred bill of fare. The diet of the grey seal varies a great deal, depending on the time of year and the waters fished. Seals fishing the offshore waters show a preference for sand lance and juvenile cod in winter, and silver hake, juvenile cod, and squid in summer. Herring and juvenile cod are devoured by grey seals in the inshore waters of eastern Nova Scotia in summer, while mackerel and squid are consumed in early winter. Each grey seal has the potential to eat over 271 kg (600 lbs) of cod a year; therefore, the population of 144,000 animals has the ability to take about 40,000 tonnes of cod. Compare this with a commercial catch by fishermen of only 10,000 tonnes in 1993, and the impact these mammals have on groundfish stocks already suffering population decline becomes undeniable.

The harbour seal occupies much the same range as the grey seal. Sable Island is a popular breeding site, where scientists have carried out considerable research. Calculations show the harbour seal population on Sable Island, perhaps typical of its range, has grown about 6 per cent, annually, over the past fifteen years, and on that basis the total population is estimated to be between 15,000 and 20,000 seals. The diet of harbour seals differs only slightly from that of the grey seal, and like all seals, they prefer soft-bodied fish, such as herring and juvenile cod.

All three kinds of seals—harp, grey, and harbour—are notorious for ripping the soft belly parts out of adult cod, fatally wounding the fish, yet wasting much of the edible flesh. It seems they have no interest in the bony head and backbone of cod, so there is a possibility that the actual damage to groundfish populations may be underestimated by the simple formula, which suggests that seals satisfy their appetite for fish by consuming the entire bodies of their prey. For a more accurate estimate of how many groundfish seals are consuming, the formula may have to be adjusted to account for this tendency to consume only the underbellies. There is no doubt that the decline of cod stocks may be at least partially explained by the increase in seal populations, their competition with cod for food, and their consumption of juvenile and adult members of the species.

Foraging by seals on the food supply of cod may also explain the recent phenomenon of weight loss by codfish. Some stocks show weight-at-age of cod to be reduced by half. One of the most severely affected cod stocks is found off eastern Nova Scotia, where the average seven-year-old cod in 1992 weighed 1.8 kg (4 lbs), slightly more than half of the 3.5 kg (7.7 lbs) that an average fish weighed in 1977. Fishermen fishing stocks with such drastic weight reduction would have to double the number of fish they catch in 1995 to land the same weight they did in 1977.

This phenomenon of weight loss has also been noted in northern cod off the southern coast of Newfoundland and cod in the Gulf of St. Lawrence. Scientists say the weight loss does not fit the usual pattern. Past experience shows that fish generally lose weight when they are *abundant* and competition for food supply is keen. When stocks are in *decline,* as they are now, average weight usually increases because the fish have less competition from their own kind in search of the food supply. However, the unusual weight loss could be caused by the steep competition from increasing populations of other predators, such as the seals.

Seals preying on cod and the food of cod and other groundfish could be part of the explanation for the groundfish crisis, yet seals and fish coexisted for many years before the seal hunt caused any serious reduction in the number of seals. The answer to the question of what has led to the poor health of groundfish stocks lies more with an examination of the impact of the human harvest of fish food stocks, such as capelin, silver hake, mackerel, and herring. While abnormally cold ocean temperatures and a flourishing seal population are, no doubt, parts of the answer for the near extinction of northern cod and other groundfish, there are often more compelling explanations that rest with the tremendous increase in fishing effort since World War II, the destructive technologies employed by the domestic inshore and offshore fishing fleets, and the often uncontrolled harvest by the distant water fleets.

Researchers aim to work with the best information at hand and with steadily improving computer-assisted instruments. There is substantial optimism that improvements to the means of measuring and investigating physical factors, such as temperature and salinity shifts and imbalances in predator-prey relationships, will give a better understanding of their influence on biological processes of fish species. More knowledge about the effects of environmental changes on fish stocks may help scientists make sound predictions for the groundfishery; however, more attention needs to be directed toward the impact of the human influence—harvesting methods—on fish and their habitats.

Chapter Four

Widening the Divide: Who Should Be Allowed to Fish for What

While nature may shoulder some of the blame for the sorry state of groundfish stocks, the ultimate responsibility rests with the tremendous growth and resulting overcapacity of the harvesting sector—domestic and foreign—fishing the waters of the northwest Atlantic. The unbridled expansion of fishing power involves an increase in the number and size of vessels that pursue cod and other groundfish, as well as improvements in the industry's ability to search, chase, and catch its valuable prey. Modern fishing gears used by some sectors of the industry inflict further damage on the resource by capturing undersized fish and non-target species and wreaking havoc upon the delicate ecosystems which support marine life.

The fishing industry on Canada's Atlantic coast is noted for its diversity. There were 30,409 registered fishing boats on the east coast in 1988, ranging in size from a dory powered by a small outboard motor to a large factory freezer ship. Traditionally, fishing vessels have been classified in one of two categories: inshore (less than 30.5 m or 100 ft. in length) and offshore (30.5 m or 100 ft. or more in length). This classification is arbitrary, disregarding important differences and similarities within the two categories. For example, inshore fixed gear fishermen, using passive fishing gears such as handlines, longlines, traps, and gillnets, have little in common with inshore dragger fishermen who actively pursue their prey with powerful net-pulling craft. Although it is an inshore vessel, the inshore dragger is simply a scaled-down version of the deep-sea trawler used in the offshore fishery—both use the same technology and method of fishing. Many inshore fishermen are involved in the day boat fishery, returning each night with their catch. The larger inshore boats in the middle distance, or midshore, fleet (20 m-30.5 m or 65 ft.-100 ft. in length) may fish what has been traditionally thought of as the offshore grounds, returning to port after several days at sea. At the same time, offshore vessels have never felt constrained from the near-shore waters. The terms inshore, midshore, and offshore do not accurately describe the location of the fishing effort but rather describe ownership of the fleet, and generally the size or classification of the vessels in a sector.

The inshore sector dominated the Atlantic fishery with about 95 per cent of those employed in the industry, before the groundfish moratorium in 1992. The inshore sector has been a key player in Atlantic Canada's economy and the lifeblood of more than thirteen hundred small coastal communities. The present crisis in the groundfishery has had a crippling impact on these small towns and villages, where the real pain of industry downturn has been felt.

Communities dependent upon the offshore fishery have also suffered from industry decline and closure. However, the offshore sector stands in sharp contrast to the inshore because of its organization. While the inshore consists of thousands of independently owned and operated vessels fishing out of hundreds of ports along the coast and sells to independently owned and operated processing facilities, the offshore is a vertically integrated operation. Three major companies rule the Atlantic Canadian offshore—National Sea Products, Fishery Products International, and Clearwater Fine Foods. The main harvesting component of this operation is corporately owned trawlers, crewed by company employees. The fish are landed for a preset price and processed by company employees at ultra-modern plants. The finished product, often ready for the consumer's table, is advertised and packaged under the company name. The entire operation moves from one step in the process to the next within the same company and is thereby vertically integrated and controlled. Before the significant decline in stocks in the late 1980s, these plants operated on a year-round basis. Since then, a number of these company plants have been moth-balled and a significant portion of the trawler fleet has been sold.

In 1992, there were 27,224 boats in the inshore classification throughout Atlantic Canada. About 16,500 of these had an active history in the groundfishery. These boats carried a small crew of between one and five fishermen and most often operated fixed gear such as traps, nets, longlines, and handlines, though a significant portion of the catching power in some areas is found on vessels that use dragger technology. One such area is the Scotia-Fundy region, which stretches from the northern tip of Cape Breton Island to the Maine-New Brunswick border. In 1992 there were more than 2,700 vessels licensed for groundfish in this region with all but a dozen classified as inshore. Of those in the inshore category, there were about 400 draggers, with the remaining 2,300 using fixed gear. About 20 per cent of the inshore fleet (600 vessels—about half of them draggers and half longliners) have generally taken 80 per cent of the annual catch of groundfish for the Scotia-Fundy region.

The handline—perhaps the least invasive gear type—is also the most popular for many inshore fixed gear fishermen working the shallower waters of the Scotian Shelf along the coast of Nova Scotia and New Brunswick. In 1988 there were 1,252 vessels in the Scotia-Fundy region engaged in the handline fishery. This fishery has only recently come under regulation and licensing on the Atlantic coast in recognition of its unobtrusive nature. With stocks in decline and the large number of fishermen using this method, some argued that an increase in take by this fleet should not be allowed.

The handline is a simple apparatus with a long history. It usually consists of a single line with one or more baited or lured hooks attached. Automation allows a number of lines to be worked simultaneously, though this is frowned upon by traditionalists. What is environmentally sound about this method is that fish caught on a handline must be craving a snack and must be a sufficient size to take the hidden hook. Experience teaches that spawning stock do not bite a hook, so there is no danger that handlines will have an impact on these seasonal aggregations, which are preoccupied with procreating. The handliner, like other fishermen, aims to maximize the value of his catch and return the greatest reward for his effort, so he looks to catch larger fish that will garner the highest price. With this motivation, he works the fishing grounds which have historically produced the largest fish and the most valuable species. He is most conscious of the seasons, the influence of the tides, and the feeding patterns of the cod, haddock, pollock, and halibut he chases.

There is no evidence to suggest that a handline fishery causes significant damage to benthic, or bottom, life forms. The hooks are lowered to a depth above the bottom; any entanglement on the

Handlining involves hook and line gear and is one of the most benign fishing methods used today. This illustration, based on old prints, shows the early version. The principle today remains the same; hooks are used to attract the fish.

ground is accidental. Generally, only the targeted species or its cousins are landed. If unwanted or undersized fish are caught by handliners, they are not usually wasted. The pressure changes felt by the fish's body as it is pulled to the surface on a handline will most often be harmless to internal organs, and the action of the hook, if confined to the hard tissue of the mouth area, will not cause serious injury. Thus, the leisurely pace of the handliner's fishing effort allows most of the by-catch to be successfully returned to the ocean.

The longline fishery, different from the handline, has a rather long history. It was first employed by the French in the nineteenth century and then adopted by vessels fishing the grounds of Atlantic Canada's coast. With the exception of a handful of boats (only two in the Scotia-Fundy region, for example) the present day longliner fleet consists of inshore vessels (under 30.5 m or 100 ft.) in length. The longlining season, unencumbered by bad weather or imposed management measures, has traditionally commenced in January and extended to the end of the calendar year.

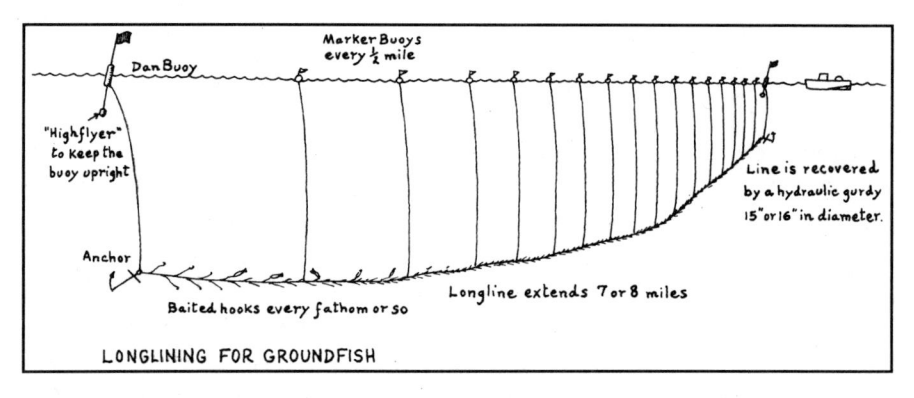

LONGLINING FOR GROUNDFISH

Labels: Dan Buoy; "Highflyer" to keep the buoy upright; Marker Buoys every ½ mile; Anchor; Baited hooks every fathom or so; Longline extends 7 or 8 miles; Line is recovered by a hydraulic gurdy 15" or 16" in diameter.

Longlining is the next logical step from handlining. Shorter lines dangle from the longline, which is anchored to lay on the bottom where groundfish mostly reside.

The longline fishery is a natural progression beyond the handline method, and its impact on the marine environment is also quite benign. The method involves baiting a number of tubs of trawl each consisting of a longline from which shorter lines dangle every 2m to 4m (6 ft. to 12 ft.). The hooks to be baited are attached to the shorter lines. Setting the longline involves casting an attached anchor and feeding the trawl through a chute and over the ship's stern as the vessel moves forward. The trawl is anchored so that it will lay on the bottom, where it will attract the targeted commercial species of cod and haddock. Larger sized hooks and bait pieces increase the chances that larger codfish or haddock will be caught. After an hour or two, the line is ready to be hauled back. The radar reflector mounted on a bobbing "high flyer" that marks the end of the trawl is located, and the line is hoisted on board, where the catch is removed from the hooks. Typically, the catch is dressed on deck in a designated area called the "slaughterhouse" and then iced below deck.

At certain times of the year, and depending on their success with the groundfishery, some longline fishermen target the prized halibut. This necessitates a change to heavier gear with stronger and larger hooks.

In its heyday, the longline fishery used to involve rather lengthy trips, from five days to two weeks. This time was necessary to reach the grounds and to fish enough gear to make a profit. Of course, larger and sturdier boats were required than those used in the day boat fishery of handliners and netters. It also involved considerable amounts of gear with crews of up to five fishermen setting and hauling as much as fifty tubs of trawl. The gear stretched for kilometres, often beyond the horizon. As catches dropped off through the 1970s and 1980s, more and more gear was required and automatic baiters became more popular in the attempt to maintain catch levels.

Most would agree that longlining is a passive, yet selective method of fishing. It requires the fish to pursue the hook, rather than being pursued with a net, and generally, longlining takes only the larger, more mature fish of a stock rather than loading the fish holds with stock of all ages. The by-catch of other species is also kept within reasonable limits. It is the opinion of scientists and most fishermen that longlining will not damage a fish stock. However, technological improvements in this method on the international stage threaten to contradict this opinion. For example, the Mustad of Norway Autoline Longlining Systems are revolutionizing this method of fishing, making it possible to equip entire fleets with a containerized longlining system. This automatic baiting, hook setting, and retrieval system is a quantum leap for the longline method, opening new grounds to those able to engage in such a capital intensive fishing effort. The result is the potential for applying more and more pressure to fish stocks.

The cod trap, another form of fixed gear, has been the primary fishing gear of the Newfoundland inshore sector for more than 120 years. Although it has undergone some modernization, the trap has remained a simple device. The body of the trap is shaped like a box and is open at the top. A lead curtain of mesh is set to run perpendicular to the current and runs out to sea from the land, where it is positioned to guide the swimming fish into the trap. Though some escape the trap at its entrance, those that enter the body become lost because instinct does not permit them to backtrack.

Before the northern cod moratorium in 1992, there were as many as five thousand Newfoundland fishermen involved in the summer trap fishery. They generally used boats less than 11 m (35 ft.) in length to work the traps, though some vessels up to 20 m (65 ft.) were used. It had become a spring tradition for fishermen to draw for berths or trap sites, each of which carried a colourful local name. The catch history of each berth was well known, so it was a matter of pride to work hard in order to maintain a site's reputation. Trap fishermen are a versatile lot, taking part in the capelin fishery in mid-June. They reset the trap using a smaller mesh. Some trap fishermen also worked in the handline, gillnetting, and longline fisheries, changing gear with the seasons.

Since the cod trap is a passive gear set to intercept migrating schools of fish, it is not a threat to benthic life, unless a significantly large catch happens to be lost during the process of transferring it to the awaiting service vessels. The dead fish could then effectively smother and litter the bottom in the vicinity of the trap. However, the impact would be short-lived and localized as scavenging species would quickly clean up the spill.

Cod traps have traditionally been constructed with 9 cm (3.5 in.) mesh; the lead curtain is made with 18 cm (7 in.) mesh. Studies have shown, however, that a large percentage of undersized fish are caught in cod traps, a concern registered by Dunne in his "Report of the Implementation Task Force on Northern Cod." Between 60 per cent and 80 per cent of fish caught in traps prior to the moratorium were shown to be under the minimum size of 41 cm (16 in.). These are fish that have not yet spawned, and with stocks in decline, a continuing cod trap fishery would have contributed significantly to the demise of the stock. A simple, though costly remedy would have required trap fishermen to convert their traps to 10 cm (4 in.) mesh. With a larger mesh, undersized fish would be able to escape from the body of the trap. The Dunne Report recognized that the conversion would be costly and the changes would result in revenue losses due to the decreases of small fish in the catches. Dunne's 1990 report recommended that the Department of Fisheries and Oceans immediately implement a gear conversion program for cod trap fishermen which would "cover all gear conversion costs" in order to encourage fishermen to make the necessary changes, including the larger size mesh. The conversion program was pre-empted by the announcement in July 1992 of the northern cod moratorium.

The most controversial fixed gear method employed on the Atlantic coast is the gillnet. Gillnet fishermen point out that the gear should pose little or no problem for target and non-target species when fished properly. Gillnets are very simple in design. Each one is a 91 m (300 ft.) curtain of mesh, anchored at both ends when set, so that it stands a short distance from the ocean bottom. The minimum legal size of the mesh is 14 cm (5.5 in.). The groundfish become entangled in the net, forcing their heads through the openings. The fish are then choked about the gills with the result that they drown. All that remains is for a fisherman to return to the net and remove his catch.

The efficiency and popularity grew among users of gillnets with improvements in the material used to make the net. Originally, gillnets were made from cotton and hemp twine, but these materials were eventually replaced by nylon and ultimately monofilament. The latter is virtually transparent and indestructible, making it the ideal material for the gillnet fishery. These same qualities make the gillnet the greatest threat to both target and non-target species.

This illustration of gillnetting shows how the net sits in the water to capture the fish.

The almost invisible monofilament of a gillnet, shown here, is one reason for the concern over ghost fishing; monofilament can last for years in the ocean.

While it is generally agreed that gillnets can be safely fished on soft bottom and in areas with weak tides, opponents to the method are adamant that they should not be fished in deeper water or over hard bottom. Hard bottom is rocky, often with treelike corals and other protrusions. Gillnets are sometimes lost when not properly tended or when torn by passing vessels. Lost nets, or "ghost nets," often become entangled on these corals and will then continue to fish. A ghost net will capture more and more fish, crab, and lobster, until the weight of the catch forces it back to the bottom. There sand fleas and other scavengers feast on the netted catch, with the result that the net becomes lighter and more buoyant so it drifts with the tides and currents. The cycle begins again as the net captures more groundfish and shellfish. It is estimated that ghost nets can continue to fish for more than five years.

The long life of ghost nets is a significant flaw of the gillnet fishery, given that the chances of ghost fishing increase with the number of nets set. While some fishermen use reason and exercise caution when setting gillnets, others display total disregard for the stocks and their supporting ecosystem. With no limit imposed on the amount of gear which can be fished, gillnet fishermen have stepped up their efforts, often in areas which pose a higher risk for the loss of nets, such as Georges Bank, which is known for its strong tides and current and areas of hard bottom. Often more than 100 nets, and sometimes as many as 1,000, are set by one crew. Between 40 and 50 nets are tied in a row, with the curtain of mesh reaching 5 km (3 mi.) or more.

Gillnet fishing captures indiscriminately. Perhaps the worst strike against gillnetting is the high discard rates, of both target and non-target species. Fish drowned in the gillnet decompose quickly. Unless regularly tended, prized groundfish become bleached and unsaleable. Consequently, they must be dumped over the side.

Although the numbers employed in the inshore fixed gear sector are greater, their fishing prowess is no match for the inshore dragger fleet. This imbalance is reflected in the share of the catch designated for the dragger fleet by fisheries managers. In the Scotia-Fundy region, for example, only slightly more than half of the inshore groundfish quota has been traditionally assigned to the 2,300-vessel fixed gear sector, while the 400-member dragger fleet grabbed a disproportionate amount, equal to slightly less than half of the grand total. Modern dragger technology is much more powerful than fixed gears in the harvest of groundfish and can catch more with greater efficiency.

Draggers and their larger and more powerful cousins, the trawlers, share a common lineage. Even today, they are fundamentally indistinguishable with regard to their use of gear. Both are termed mobile gear because they actively pursue their catch, chasing and scooping it from the bottom. Both draggers and trawlers have been blamed for causing considerable damage to the ecosystem of inshore and offshore fish stocks because they plough the ocean bottom to fill their nets.

Fishermen have fished with nets for thousands of years and continue to do so today, with tremendous technological improvements over the past two decades, advancing harvest ability by light years. All sizes of vessels sailing the world's oceans have adopted ultra-modern equipment and huge nets, demonstrating a new-found harvesting efficiency. A fisherman no longer has to rely on his experience and wits for fishfinding. Rather, he can power up his computer-assisted electronic wheelhouse. He can place his vessel on automatic pilot, setting a course for the fishing grounds; locate the fish on a monitor; deploy the fishing gear, adjusting its location to capture the catch; and then return to port. Though there are still variables in fishing, including weather, price, and supply, more and more of the uncertainty has been removed from the fishing effort.

The first significant improvements in the mechanization and modernization of fishing effort came in the 1800s with the introduction of steam and engine power. Prior to the nineteenth century, the industry had depended on the power of the sail. The new energy sources increased the mobility and catching efficiency of fishing vessels as well as the on-land transportation of fish—a perishable product—to market.

The forerunner of the modern dragger and trawler technology—the otter trawl—benefited most from mechanization. The trawl gear used a woven mesh net to capture the fish and was towed by sail-powered boats. It was first developed in Europe, although a variation of this gear was also independently brought on stream by Japanese fishermen. The first trawls were simple in construction. Refinement of the method resulted in the development of the beam trawl, whereby the net was attached to an oak beam of about 4 m (12 ft.) in length. The beam was used to keep the mouth of the trawl open, even when the vessel was motionless. This was an important innovation at the time because sailing vessels were often becalmed, and the mouth of the net would then close. In addition to the placement of a beam across the mouth of the net, runners were attached at either end of the beam to keep the net off the bottom, reducing resistance from the seafloor. This North Sea trawling method became more popular. Augmented by the new sources of energy—steam and engine power—it resulted in an expansion of fishing power. Steam-powered and engine-driven trawlers were predominant, reporting catches six to eight times greater than that of sailing vessels. The first successful prototype of the present day otter trawl was developed in 1892 in Granton, Scotland. The tremendous improvement in catching efficiency with the otter trawl was so quickly appreciated that it became universally accepted and adopted.

The otter trawl offered a number of distinct advantages over mobile gear already in use. Otter boards or doors are attached to the net to serve the same function as the rigid beam of the beam trawl. The wooden doors, when towed, respond to the increased water pressure by spreading apart, forcing the net open. A metal strike plate or bank of flat iron is attached to the shoe, or bottom edge of the door, to slow the inevitable wear from dragging along the ocean bottom. The net has a mouth stretched wide by the attached doors and a footgear comprised of metal bobbins, or rollers, which form the "jaw" or lower portion of the net that opens in near contact with the seafloor. A headline fitted with floats lifts the upper lip of the net, giving height to the opening. The body of the net is funnel-shaped with sidewalls narrowing to a bag called the codend, where the fish become trapped, unable to escape through the mesh. The end of the net is called the codend because of its shape; the fish we now know as codfish acquired the name through association with the gear that specialized in its capture. "Cod" was the Middle English word for "bag." Thus, codfish was the name given to the most common fish captured by the trawl.

With the introduction of a steam winch and a double drum on deck, the new trawl proved easier to handle and manoeuvre and also easier to haul. The otter trawl also allowed for a much larger net. By 1895 the entire British trawler fleet had adopted the otter trawl, and its use spread to neighbouring countries fishing the North Sea.

Groundfish in the North Sea came under increased pressure in the years leading up to World War I. Reduced catches off the British coast had forced the fleet to look further afield in order to maintain production. The outbreak of wartime hostilities made the entire northeast Atlantic unsafe for fishing enterprises, giving a much-needed reprieve to fish stocks. By 1918 groundfish stocks were the healthiest and most abundant they had been for a half-century. The recovery clearly showed the devastating impact recent trawling efforts had on North Sea stocks and should have signalled those concerned with the maintenance of a healthy fishery. The indications went unheeded, however, as haddock and flatfish stocks again came under intense pressure from a retooled postwar fleet determined to meet the growing market for fresh product for the fish-and-chip trade.

The European release of the otter trawl was soon followed by its introduction in North America. The first trawl was loaned to a Cape Cod fisherman as early as 1893 by the United States Fish Commission. The experiment proved too much of a temptation and its use was quickly adopted by New England fishermen who were impressed with its efficiency. The American fishery, based on the free enterprise system, was ripe for the introduction of the otter trawl much more so than its more conservative and traditionalist counterparts to the north.

Efforts to establish the use of the otter trawl in Canada met with stiff opposition. An industry-wide protest during the 1920s resulted in the appointment of a federal Royal Commission to investigate the appropriateness of dragger technology. The 1928 Royal Commission Investigating the Fisheries recommended that further development of the dragger fleet in Canada be put on hold. The Canadian federal government accepted the recommendation, forbidding additional expansion. It was not until after World War II that the push to industrialize the fisheries proved too great a force, and former resistance was cast aside, allowing the Canadian fisheries to embrace the newest innovations in dragger technology.

With the end of World War II, the Canadian fishing industry busied itself with gearing up for a new age. The domestic industry adopted many of the modern developments produced by other fishing nations. New technologies devoted to finding, catching, and handling fish increased the power of the Canadian trawler and inshore dragger fleets. This expansion was spurred by the ever-increasing competition from a European distant water fleet which pursued stocks up to the Canadian coastline. Not only did the harvesting sector expand to compete, but the shore-based, or processing, sector of the industry also kept pace by steadily improving its performance and output.

One of the most significant Canadian developments in fishing gear was the Atlantic Western trawl, or simply Western trawl. Trawls in use on the Atlantic coast prior to 1960 had been constructed with two panels, a design that was European in origin. Development by gear technologists involved in work on the Pacific coast had resulted in a four-panel design that provided for a greater vertical opening and much better catches of high-swimming species like haddock and cod. Another advantage was that the four-panel trawl was designed to fish with less netting near the ocean bed, making it less vulnerable to wear and damage. Also, the use of very fine twines and larger meshes in the wings of the gear made it possible to tow the box-shaped trawl with much less power. Further improvements to the four-panel Western trawl to suit it to Atlantic coast conditions made it a powerful tool for trawler and dragger fleets. The new four-panel design, augmented by the Atlantic development of hydraulic-driven trawl reels or drums, immediately translated into greater catches of cod, haddock, and pollock.

While adoption of the Western trawl greatly enhanced the catching efficiency over the otter trawl, the manner in which it was deployed was also the focus of innovation. Prior to World War II, the most popular method of trawling involved the setting, towing, and hauling of the net from the side of the vessel. These side trawlers, as they were called, towed the net from the gallows—a massive, well-braced, steel structure anchored to the deck—and from a towing block or pulley located on that side. In a traditional side trawler the wheelhouse, or pilothouse, was located at the stern of the vessel, with the engine room below.

During the 1950s, a new design emerged on the Pacific coast of North America, whereby the trawl gear was set and hauled over the stern. On stern trawlers, the wheelhouse, or superstructure and living quarters, were located forward, clearing the way for more efficient use of work space at the stern of the vessel. Stern trawlers offered distinct advantages over side trawlers. With the gear worked over the stern, the ships could maintain a straight course, avoiding the need to sail across wind, so it was possible to work in rough sea

A large trawler's Western trawl—an innovation from the Pacific coast trawls. The Western trawl has four panels to allow for a greater vertical opening and uses less netting near the ocean floor.

conditions. Also, the working deck at the stern is sheltered by the forward positioned wheelhouse, and generally the boat is much more stable, less susceptible to pitching and rolling. The stern trawling method, first adopted in the 1950s by British Columbia fishermen, was in vogue on the east coast by the early 1960s. Because of the increased efficiency of the stern trawling method, side trawling on both coasts was quickly displaced.

The inshore dragger fleet has made significant improvements to its vessels. The most recent improvement in design was made possible by the industry's acceptance of fibreglass as a material in the construction of a ship's hull. Vessels prior to the 1980s had been limited in their size and range of construction. The flexibility and strength of fibreglass changed that. While conventional wooden-hulled vessels measuring over 13 m (44 ft.) in length had a carrying capacity of 23 tonnes, the new

Hydraulic-driven trawl reels or drums allow for even greater fish catching capacity.

The trawl is towed by the gallows—a massive steel structure anchored to the deck—on a side trawler, which were mostly supplanted by stern trawlers.

deeper and wider "jumbos," constructed with fibreglass hulls and carrying a much wider beam, were capable of floating as much as 164 tonnes. The jumbos, whose construction abandoned the sleeker lines of the traditional Cape Islander, were shaped more like bathtubs and featured greater catch carrying capacity, as well as nearly all the luxuries of home: full bath, stereo music, television, and rather spacious living and sleeping quarters. The price tag reflected the built-in luxury, jumping from the $200,000 price of conventional draggers to about $750,000.

The approach to midwater trawling has also changed to achieve greater catches and better efficiency. Following World War II, considerable effort was expended to develop the Canadian midwater trawling fleet. In midwater trawling, the net is towed high in the water column and off the bottom to catch redfish and pelagic species, such as herring, that school at various levels between the ocean's floor and surface. A number of vessels were fitted to combine midwater trawling with purse seining (a fishing method that involves the pursing of a large curtain of mesh which is set as the vessel travels in a circular path to surround the catch), when the pressure on herring stocks gained momentum during the 1960s on both the Atlantic and Pacific coasts. It was also discovered that these midwater trawls could be used to capture redfish which frequent the upper portions of the water column at night. This groundfish species then became the target of both inshore and offshore mobile

Redfish are taken in midwater trawling and by foreign fleets that have fished out their cod stocks. Hauling in the very large trawls often damages the fish as they get squeezed through the mesh.

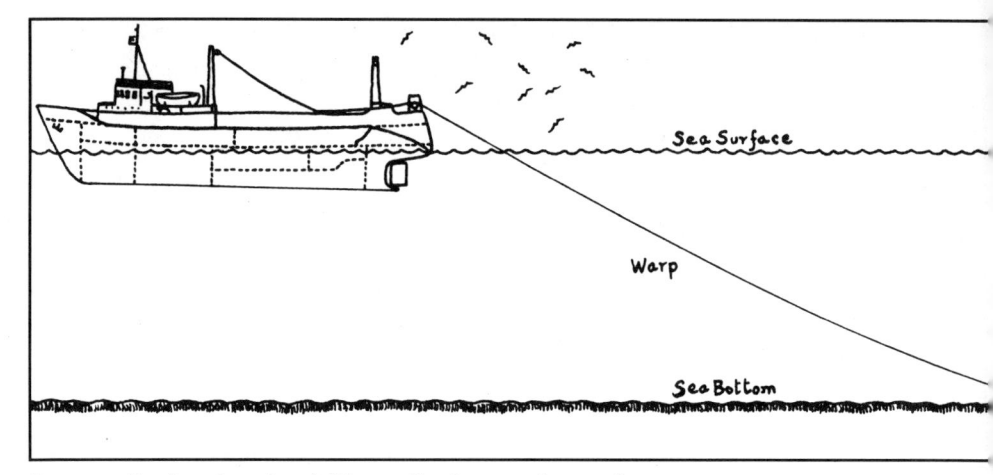

Stern trawling largely replaced side trawling because the vessels use space more efficiently, allowing fishermen to fish on rough seas, and the stern trawlers are more steady on the water.

fishing fleets working the deeper waters of the Gulf of Maine, the Gulf of St. Lawrence, and the banks off Newfoundland and Labrador.

Today's version of the midwater trawl, used by the international distant water fleet fishing the world's oceans, poses perhaps the greatest threat to those species which occupy the upper portions of the water column. The gear now consists of a trawl measuring in excess of four football fields in width and a height of 160 m (525 ft.). The mouth of a trawl is a gaping 35,800 m^2 (9 acres) that virtually vacuums up species like redfish swimming in the North Atlantic. The net is the newest version of the Gloria trawl, developed specifically to herd fish which do not swim in tight schools. The redfish, for example, has become a targeted species now that fishing fleets such as those from Iceland, Norway, the Faeroe Island, and others have fished out their cod stocks. The new trawl is gaining popularity as the Icelandic manufacturers brag that more than fifty-five of the trawls have been sold to European fleets, which will first target the ridges of deep water, southwest of Iceland.

Canadian fishermen fear that this same gear may be turned on redfish stocks on the Flemish Cap, just off the Grand Banks of Newfoundland and beyond the 200-mile limit. The vessels involved in the midsea redfish fishery are powered by engines of horsepower between 2,500 and 4,500; they expect to catch and process about 30 tonnes of fish daily. According to *Fishing News International,* the first trawls of this type had a circumference of 1,100 m (3,608 ft.), that increased in 1993 3,072 m (10,076 ft.), with a maximum mesh size of 128 m (420 ft.) for the foremost portion of the panels.

Midwater trawling also uses some of the most advanced fishfinding equipment available today. This electronic gear is used to direct the net at the targeted catch. The wheelhouse is the nerve centre of the operation, receiving

Widening the Divide

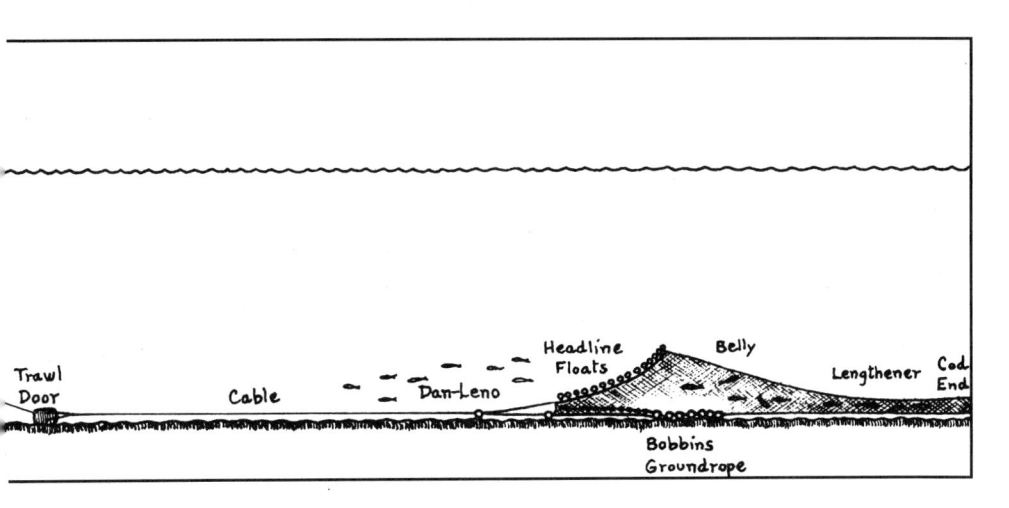

[Diagram of trawl gear labeled: Trawl Door, Cable, Dan-Leno, Headline Floats, Belly, Lengthener, Cod End, Bobbins, Groundrope]

input from sonar, echo sounders, net-mounted sensors, and navigation and propulsion systems, all assisting in the adjustment of the net's position in the water in order to capture the most fish. With deck-mounted winches or power-driven drums, the cable warp (or tow line) attached to the net is adjusted so that even when the vessel is changing course, the gear is properly positioned to maximize its take. The gear, having been directed to the best depth and position, is then placed under automatic electronic control. The relative position of the net, the seabed, and the school of fish are all in the captain's view, while another reading warns when the codend is becoming too heavy with the catch. Nothing is left to chance. The catch is maximized and damage to the gear is minimized.

    Although the net of a midwater trawler is huge, its daily take is small next to the catch landed by the massive factory freezer trawlers working the northwest Atlantic. Freezer trawlers from countries including Spain, Portugal, France, and Russia measure 91 m to 137 m (300 ft. to 450 ft.) in length, with a carrying capacity of up to 4,000 tonnes. These huge vessels are driven by twin diesel engines producing in excess of 6,500 horsepower. The rugged towing and lifting gear can pull tens of tonnes of groundfish from the water with each set of the net. The Spanish fleet is particularly adept at fishing the waters off Canada's east coast with the trawlers working in pairs. Both vessels cooperate to pull a single net which may measure nearly 2.5 km (1.5 mi.) in length. Each takes a turn pulling and emptying the net to maximize so the efficiency of the harvesting effort. Below deck the fish are processed with the assistance of a network of conveyors and automatic splitting and salting equipment. Those vessels designated for the production of a frozen product are equipped with filleting and quick freeze machinery.

    The Canadian offshore trawler fleet, though successful in its own right, can hardly match the proficiency of the European distant water fleet. The Canadian offshore sector fishes from multimillion-dollar steel vessels measuring about 46 m (150 ft.) in length—less than half the size of the average foreign vessel on the outer reaches of the continental shelf. Only one vessel, National Sea's *Cape North,* measuring 82 m (270 ft.) in length with the capacity for processing and freezing product, comes close to the size and catch capacity of the heavyweights floated by the foreign fleets.

The focus of research carried out by Canadian gear technologists over the past forty years has been applied to the equipment used in bottom otter trawling. The Canadian fishing industry had been impressed by the rapid development of fish catching technology introduced after World War II. These improvements had been readily adopted by foreign fishing nations and the increased proficiency of the fleets was becoming evident, while the percentage of the domestic take of northwest Atlantic groundfish fell dramatically. Those in the industry concerned with maintaining Canada's competitiveness pushed for modernization of the domestic fleet and its fish catching capability. The primary push for improvement came from the large offshore companies that were already focusing on otter trawling as the wave of the future. Much of this research over the years has focused on modification of the groundgear and doors—that portion of the trawling gear which is in contact with the seafloor.

The groundgear, or footgear, designed to catch groundfish such as cod, haddock, and pollock have undergone revolutionary change over the years. The physical change to gear was minor, but the resulting improvement has opened up vast areas of productive ground once considered too difficult for trawls to fish. This expanded access was brought about by the simple replacement of conventional roller gear, consisting of spherical bobbing, with what has become known as Rockhopper footgear. This Rockhopper gear uses large discs, usually 30.5 cm to 36 cm (12 in. to 14 in.) in diameter, spaced at intervals with smaller discs, generally 10 cm (4 in.) in diameter. The smaller discs prevent displacement of the larger ones. When this footgear comes in contact with a rock or other obstacle, it tends to hop over it, preventing the trailing net from snagging on the rough bottom, thus reducing wear on the gear. This rough ground is, of course, a favoured shelter for groundfish hoping to escape predators, though it cannot protect them from the trawls.

The footgear is adapted to target a particular type of groundfish. For flatfish including flounder, plaice, and sole, the groundgear is preceded by a series of "tickler" chains. These disturb the muddy bottom on which the fish rest, creating clouds of dust and sand that drive the flatfish upward so they can be caught by the trailing net. The number of chains and their weight have been the focus of experimentation in local fisheries, where efforts are made to produce the greatest efficiency of catch. For cod, haddock, and pollock the Rockhopper and bobbins groundgear accomplish the same result. These footgears create the desired dust clouds and additional noise. Groundfish, which ordinarily hug the bottom and avoid capture, are frightened by the

Rockhopper footgear is designed to prevent the trawl from tangling on the ocean bottom by hopping over any obstacle it meets. This footgear is used to target groundfish such as cod and haddock.

commotion of these visual and auditory stimuli, so they move upward, into the path of the oncoming net.

Overcoming the rough rocky bottom also led to the development of otter trawl doors which would not get caught on obstacles. Since experiments have shown that the trawl doors produce as much as 30 per cent of the total drag or bottom resistance created by the gear, improvements have been made in this area too. A wide variety of shapes and sizes of otter trawl doors have been produced, with some more suitable to local conditions than others. An example of one of these innovations is the Bison trawl door. A promotional video produced by the Marine Laboratory, in Aberdeen, Scotland, and circulated by the Canadian DFO, explains why the Bison door is worth investment by dragger fishermen: "The Bison door clears obstructions with ease and what is important is that they do not lose stability or spread and immediately come back to the seabed. Ground contact with the Bison door is very good. Commercially the Bison door is proving to be very successful on all grounds, especially on rough ground, on peaks, in strong tides and in rough weather, with proven results." The video shows the Bison door overturning large boulders, uprooting anything of its own size and smaller. Clanging from rock to rock, the Bison door performed fully as well as the narrator described, returning to the seabed immediately after smashing headlong into any obstacle. No acknowledgement was made of the environmental damage inflicted by the gear.

Gear technologists are constantly working to improve a trawler's ability to fish rocky bottom, giving the distinct impression that soon no ground will be too rough or beyond reach. Dragging over rough bottom has invariably meant damage to the doors, cables, and nets, but a recent rigging trick by a French manufacturer, using vertically positioned trawl doors, may reduce damage to

Studies have shown that trawl doors produce up to 30 per cent of the total drag created by the gear on the ocean bottom. Researchers are devising doors to improve accessibility of fish in rocky areas that will keep mouths of trawls open and prevent snagging.

gear. The vertical trawl doors on the prototype fly above the seafloor while the gear stays on the bottom. Though no acoustic listening device is available to tell the skipper the door's position relative to the sea bottom, the geometry of the gear, controlled by its rigging, results in the doors maintaining a height of about 8 m (26 ft.) above the floor. The gear is arranged so that, should the doors snag, they can be lifted using the upper bridles, or cables, attached to the doors, thus limiting gear damage and shoe wear. Fish herding is accomplished with the bridles that reach from the doors to the net. The required sand cloud which scares the fish into the path of the net is created by the groundwire's counterweights—a cable hanging from the groundwire and dragging along the seafloor. The new gear has been used to open up very rough ground, making available stocks of the deep-water groundfish species grenadier. According to the *National Fisherman,* Norwegian ships also use the gear while fishing for turbot, or Greenland halibut, at depths up to 1,650 fathoms.

Research scientists conducting studies on the ocean environment have also had to contend with the destructive nature of trawls. For example, study of current and ice movements on the fishing grounds east of Newfoundland has required development of a modified trawl-resistant package made of aluminum and shaped like a pyramid with the peak sliced off. This package houses a research tool called a Doppler current profiler, which measures the velocity of water movements along the ocean floor. As described in the *Sou'Wester* of February 1, 1990, it maintains a flat position on the ocean bottom and is designed so that trawls will not root it from its position. Instead, the gear climbs up one side of the box, across the top, and down the other side.

Scallop dragging and clam dredging methods carry implications for groundfish because they scour a bottom shared by cod, haddock, and other such species. The impact of these fishing methods upon shared groundfish habitat is, in many ways, much more devastating than the effects produced by groundfish draggers and trawlers. The two sectors of the scallop industry employ a similar technology, though on a different scale, in much the same way that the groundfish dragger used by the inshore, and the groundfish trawler used by the offshore are different generations of the same beast. The scallop rakes used on the fishing grounds off the Atlantic coast are the product of many years of development. The offshore version consists of a cage with a pressure plate at the front and a collecting bag of interlocking steel rings at the rear. Powerful draggers tow two or three of these rakes, each 4m to 6 m (13 ft. to 20 ft.) in width and 650 kg to 700 kg (1,430 lbs to 1,540 lbs) in weight when empty. When full, each dredge can weigh up to 3,600 kg (7,920 lbs). The weight of these rigs is borne largely on the edges of the pressure plate, which cuts into the ocean floor. The offshore rakes also utilize sweep and tickler chains which disturb the upper layer of the floor, forcing the target species and other life into the collecting bag. The inshore scallop draggers carry one or two gangs of dredges, each employing up to seven rakes. Each rake weighs 34 kg (75 lbs) and measures 0.76 m (2.5 ft.) wide.

The hydraulic clam dredge is used on the Scotian Shelf and Grand Banks in the pursuit of hard-shelled clams. Vessels of 24 m (80 ft.) or more in length use a hydraulic pumping system to blast the clams from their homes in the upper layers of sediment on the seafloor using a pressure of 125 lbs/in^2. Water jets are forced through a series of nozzles along the front edge of the clam dredge, liquefying the sediments to a depth of 15 cm (6 in.) or more. The clams are then excavated by the cutter blade, located just behind the bar of nozzles, and drawn upward into the collecting cage at the rear of the dredge. When the dredge is ready to be retrieved from the seafloor, it is hauled from the water by the lifting cable attached to the front of the harvesting apparatus. Large vessels working deeper waters and in more threatening weather handle the heavy hydraulic dredge by pulling it up on a sloping metal frame secured to the deck of the boat. The bag of the dredge is then in a position which makes for easy unloading.

While fish catching techniques have undergone rapid and dramatic improvements, fish location has advanced at an equally rapid rate. Satellites are now being used to provide information for fishing vessels, including sea

Net Destruction

Offshore scallop draggers use cages like this to drag along the bottom. A pressure plate at the front of a collecting bag cuts into the ocean bottom, damaging groundfish habitat.

Heavy chains precede the catch cages on scallop rakes to stir the seafloor and uproot the scallops from the bottom.

surface temperatures and concentrations of plankton, both of which indicate where to locate commercial stocks. In 1994 NASA launched a satellite called *Seastar*. This eye-in-the-sky is designed to indicate to scientists changes in atmospheric gases and to indicate to fishing vessel captains locations for a big catch. The satellite pinpoints areas of phytoplankton concentration and thus the fish that feed on this floating meal. Fishermen need only dial for a direct radio link with *Seastar* and set sail.

The wheelhouse gadgetry aboard modern vessels, from factory freezer trawlers to the smallest inshore dragger, is an on-screen disclosure of the technological advances in the electronics industry and the refinement of techniques in underwater sound. Long-range sector scanning sonar equipped with automatic programs enable fishermen to better locate fish concentrations from a greater and greater distance. The most modern sonar can read the magnitude of fish concentrations, determine their range, their bearing and their movement, for as many as nine separate schools at one time. This equipment can even identify the species under observation, complete with colour patterns, and indicate the relative position of vessel, net, and fish. The blips, each representing a fish, can be displayed on a variety of colour monitors to fully coordinate the fishing effort —all of this is fully computerized. Similarly, echo sounders have kept abreast of advancement, providing high resolution images with dual readouts—one for fishfinding and one for bottom discrimination.

It may be assumed that this sophisticated equipment is found only on the largest factory freezer trawler, but this is not the case. In addition to the series of gauges sensing the engine and hydraulic systems, the typical inshore dragger working the Atlantic coast carries two radar units, two Loran C units, a wet and dry paper echo sounder, an automatic pilot, a couple of VHF radios and citizens bands, and a plotter. Much of this equipment is now integrated,

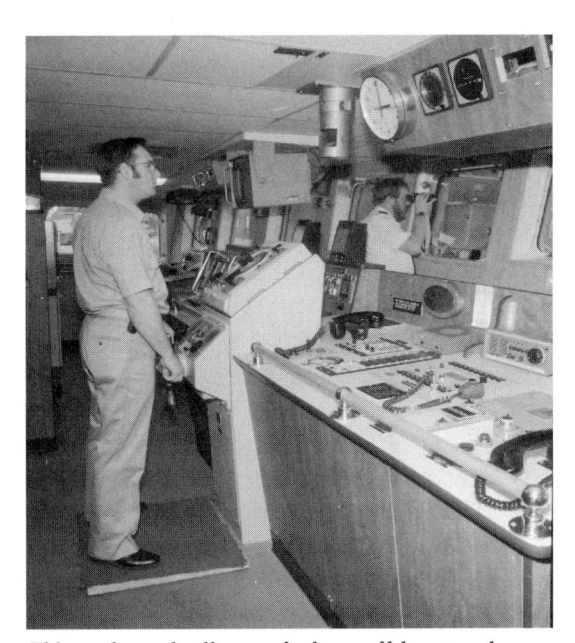

This modern wheelhouse of a large offshore trawler shows the fishfinding gadgetry that aims to increase efficiency and ease of fishfinding.

as electronics companies compete to provide fishermen with the latest advanced package. A November 1, 1993, paid commentary in the *Sou'Wester* newspaper promoted a combination GPS (global positioning system) receiver, plotter, and fishfinder. The GPS receiver contains five channels simultaneously tracking up to five satellites to provide an update on the vessel's position every second, anywhere in the world. In the plotter mode, the equipment can display the previous fishing track and the course line to be followed to the next waypoint. The fishfinder has seven depth ranges to 195 m (640 ft.), and split-screen options are available for combination displays of both fishfinding and navigational data.

The explosion in fishing power for both the inshore draggers and offshore trawlers working Canada's Atlantic coast has meant a tremendous increase in the harvesting pressure directed toward groundfish stocks. It is not only the landed catch of targeted species of market size that is subjected to this harvesting pressure. Undersized fish, fish caught beyond the allowable catch, and unwanted species are also inevitably trapped in the codend of the trawl. The net indiscriminately gathers in both large and small groundfish, spawners and juveniles, trash fish and shellfish. Only groundfish of market size and amounts caught within the quota are to be kept. The remaining huge, unmeasured, and undetermined biomass is then turfed over the side or ground into meal. The net is then prepared for another tow.

Discards of species brought on board are a corollary of intensified and indiscriminate catching capacity. Discards, meaning the amounts tossed into the ocean from the ship's deck, do not tell the entire story. Authorities are aware that some companies and independent fishermen are responsible for dumping or releasing the entire catch from a net. The discard rate for the offshore trawler fleet on the Scotian Shelf during the late 1980s was estimated at about 5 per cent of the total catch. A snapshot of dumping in

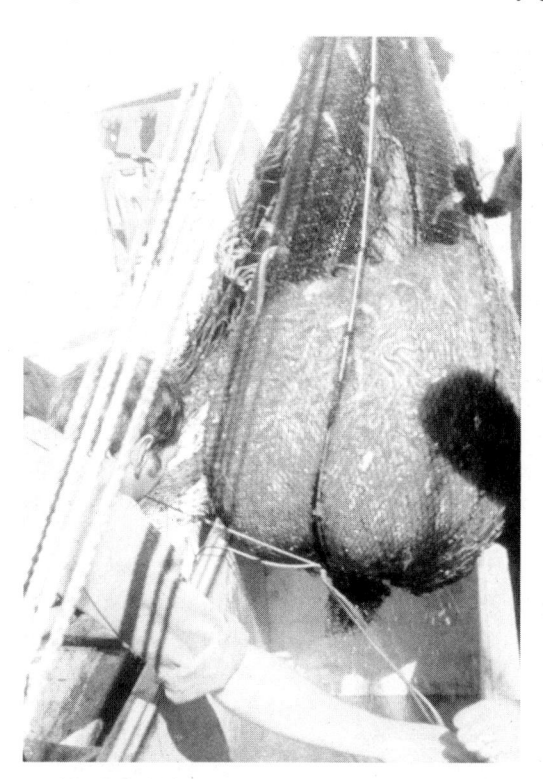

Trawls "indiscriminately gather in both large and small groundfish, spawners and juveniles, trash fish and shellfish."

Halifax's *The Chronicle-Herald* on November 26, 1992, showed that at the height of the fishery in 1989, draggers in the southern Gulf of St. Lawrence dumped more than 11 million undersized cod, or 15 per cent of the total catch. Federal scientists point out that with other stocks in the Atlantic region the rate of dumping was probably higher. According to an article in *The Chronicle-Herald* on March 10, 1990, statistics gathered by the scientific community were disputed previously in 1990 by fisheries managers, who claimed that the situation in Canadian waters was much less serious than scientists claimed.

While Canadian fisheries managers have been reluctant to admit that discards are a serious problem, the European Economic Community (EEC) has admitted that the situation is out of control. A special EEC report in 1992 notes that discards by offshore trawlers working the North Sea can be significant but also blames inshore trawlers working coastal waters. Discard rates for North Sea haddock, for instance, often exceed the amount kept for each tow. The 1985 figures indicated that 460 million haddock were discarded with only slightly more—500 million—landed. Discards are a problem worldwide. Norway has taken the most direct approach to curb the problem by banning discards, though such a regulation is nearly impossible to enforce.

Gear technologists have expended considerable time and energy in efforts to reduce discarding. The focus has been on developing new types of meshes and on adjusting minimum mesh sizes to allow the escape of juvenile fish from dragger nets. However, the problem of the waste of discards is not so easily remedied. While varying survival rates for fish escaping trawl gear have been observed, there is growing concern that the escapees suffer appreciable damage, which places them at risk after the fishing vessel has moved on. Scottish studies by Main and Sangster indicate that more than 80 per cent of haddock escaping from inside the trawl die. Soviet scientists Zaferman and Serebrov used an underwater submersible to observe appreciable numbers of dead cod and haddock on the seafloor along the path of a trawl immediately after its passage. Norwegian studies by Soldal et al. do not agree with these findings and suggest that mortality of cod is negligible and of haddock is less than 10 per cent.

It has been proven, though, that flatfish are especially vulnerable to injuries inflicted by trawls, which uproot them from their place on the seafloor. Serious harm in scale loss and bruising is also caused when the flatfish make their way through the meshes of the trawl resulting in scale loss and bruising. For example, the Beck et al. survey reported that the survival of sole that escape through meshes is estimated at only 60 per cent. In the commercial beam trawl fishery of the North Sea, the survival of both plaice and sole discards from the entire fishing operation, including on-deck handling, is estimated at less than 10 per cent. The catching process involves the use of a series of heavy tickler chains and these, coupled with injuries caused by crushing and jamming in the net, inflict the greatest damage. Given the rough treatment during the fishing process, it hardly matters that the sorting of the catch on the ship's deck also claims its share of casualties. A 1989 Canadian study by Neilson et al. showed that only 35 per cent of small halibut brought to the surface after being fishing by trawl survived, while 77 per cent of those from a longline catch survived after being returned to the water. Herring and mackerel are most vulnerable to the capture process. In a 1993 study conducted by Suuroven et al., it was found that small herring escaping from a 36 mm diamond mesh codend used in the Baltic Sea showed a survival rate of only 10 per cent to 15 per cent.

A closer examination of the fishing of otter trawls shows that both target and non-target species suffer damage from the equipment used. The fish are herded by the gear; the doors of the otter trawl are the first visual stimulus detected by the fish. The gear moves in on the catch, and larger fish, such as cod and haddock, respond by swimming for long periods at the mouth of the trawl. Most fish species can maintain a speed of 2 knots, but at greater speeds the cod or haddock begin to tire after approximately ten minutes and fall back into the funnel created by the doors and the resulting sand clouds.

The damage to fish in the process of trawling further increases fish mortality, although there is no single reason for the death of the fish. Some escape through the front panels of the net, but most reach the codend, where they come in contact either with the meshes or the building wall of the catch. If small enough in girth, fish either actively swim through the meshes or are forced out by the turbulence created by the moving trawl. Those fish that do escape give a number of rapid tail beats to break free and it is at this moment that the fish appear to suffer the most of the scale loss. Scientists have noted a distinct billowing of fish scales around the codend at the time fish were actively involved in squeezing through the meshes. Fish spend varying amounts of time in the codend before escaping. In the codend, they are damaged by being forced against the meshes, by rough-skinned and horned species like dogfish and shellfish, and by trash scooped from the bottom, including rocks and other debris. In addition, muscular fatigue and exhaustion have been shown to take a toll. This is most pronounced in haddock—up to 27 per cent of the mortality from trawling has been attributed to muscular fatigue according to an article by Beamish that appears in a publication titled *Fisheries Research Board of Canada*. Cod, on the other hand, do not appear to die as the result of physical exhaustion.

Studies have given some recognition to losses beyond those of recorded catches, but it would seem appropriate that a factor be entered for mortality of escapees from the trawl net. Harris, in his 1990 "Independent Review of the State of Northern Cod Stock," reported a loss of northern cod equal to 30,000 tonnes a year to account for discarded by-catches of fishing effort

Landing herring. Herring are one of the most vulnerable species to the capture process because the smaller fish that do escape the net show the least probability of surviving the damage and/or trauma.

directed for other species, discards in excess of legal limits by both domestic and foreign fleets, discards of spoiled fish from poorly attended gillnets, and fish caught in ghost nets.

There has been long and steady opposition to bottom trawling dating back to the Renaissance. Competing fishermen working the coast of England and the Netherlands, soon recognized the great threat posed by the new gear termed *wondyrchoun* or *wonderkuil,* because it involved towing a small meshed net between two vessels. They wasted little time in pressuring the authorities to limit or even ban its use. The earliest concerns were surprisingly well informed and well articulated and, for a time, were effective in controlling the use of bottom trawlers.

Originally, the opposition was focused on the small fish being trapped in the fine meshes of the trawl. This prompted the Parliament of England as long ago as 1350 and again in 1371 to pass legislation banning the use of trawls. Concern was also expressed for the ocean bottom upon which the trawl was pulled. A petition to the Parliament of England in 1376-1377, translated from the original Norman French of the day and quoted by De Groot in 1984, cites concerns for the plight of undersized fish and for the benthos, or ocean bottom: "...To which instrument is attached a net of so small a mesh that no kind of fish, however small, that enters it can pass out, but is forced to remain within it and be taken. And besides this, the great long iron of *Wondyrchoun* [trawl] presses so hard on the ground when fishing that it destroys the living slime and the plants growing on the bottom under the water, and also the spat [larvae] of oysters, mussels and other fish, by which the large fish are accustomed to live and be nourished."

Recent scientific investigations have proven that fishermen's early concerns about trawling were well founded. The justified concern did not dissipate, although its influence was overridden by temptation, which led to greed and greed to denial as the authorities and those with influence within the fishing industry chose to ignore the cautions in the pursuit of greater landings. The rush to harvest the riches of the sea made more and more accessible by industrialization and mechanization of the industry meant that concerns for the impact on the ocean floor and on fish populations took a back seat.

It was not until 1970 that the International Council for the Exploration of the Sea (ICES) received and adopted a resolution calling for study of the effects of trawls and dredges on the sea bed. The modern age of investigation was then ushered in. French delegates to the meeting pointed out that a ban on beam trawling in the Mediterranean had already been granted after it was discovered that the heavy gear had significantly altered the sea bottom near the island of Corsica. France was concerned that a similar impact would result from the intensive trawling conducted by Dutch and Belgian trawlers along its Atlantic coast. The beam trawlers, fishing in these waters for Dover sole, had stepped up their effort and were using heavier gear. The original single tickler chain had been replaced by a gear consisting of between six and sixteen chains weighing more than 3 tonnes.

The effects of beam trawling, otter trawling, and the dredging for shellfish had created a number of concerns, to which the 1970 ICES meeting was determined to respond. As well as anxiety over the aggressive depletion of stocks with loss of young fish trying to escape the trawl, concern was also expressed that other life forms were being disturbed or destroyed. It was feared this would lead to a loss of food supply for commercial stocks. Also, the heavy gear was shown to roll boulders and other objects from their partially buried positions to the floor surface, where they posed an obstacle for lighter bottom trawling gears. Some researchers advanced theories that countered perceptions of inflicted damage. The initial reaction of a number of researchers was that the damage caused was minimal, even insignificant, when compared with natural causes, mortality, or the influence of wind, wave, and tide. Some even suggested that the surface scouring by the heavy tickler chains used in beam trawling and the footgear of otter trawls would have a beneficial effect as beds of mature life forms were thinned, allowing for new growth.

The remaining forms, either dead or damaged, would provide a meal for the prized commercial stocks.

Concern for the damaging effects of trawling persisted. Unwilling to be turned back or distracted, the ICES embarked in 1970 upon an exhaustive study of the impact of this gear on the ocean bottom. With refinement in measuring techniques, the approach of investigation has become more and more sophisticated. In addition, momentum has grown over the past twenty-five years as ICES researchers from both sides of the Atlantic have joined efforts to study the impacts of trawls. In fact, researchers around the world have been documenting the impact of trawls.

The ICES Gear and Behaviour Committee was breaking new ground in 1970 as their initial studies attempted to scientifically measure the impact of bottom trawls. Underwater investigative techniques in use during the 1970s would now be considered rather primitive; these earliest probes were limited. Earlier studies did not show the same degree of negative impact as more recent research does because the work of the seventies was carried out on light gears towed by less powerful trawlers and at low speed. The trawlers of today are huge by comparison, are equipped with heavier gears, and utilize horsepower undreamed of only a decade ago. Today draggers sweep the ocean bottom with a force up to 10,000 kg (22,000 lbs) of pulling power. Also, the earlier experiments revealed the effect of a single pass by the trawl while studies show that some portions of fishing grounds are trawled as frequently as ten times a year.

One of the very first scientific studies was carried out in 1947 by Ketchen of the federal Pacific Biological Station in British Columbia. The study involved laying out a course on a tidal flat to be dragged during high tide. The bottom, of intermediate hardness, was scraped and the eel grass was sheared but not uprooted by the trawl doors. Ketchen surmised that the damage would have been more severe on muddy bottoms and less on hard sand or gravel bottoms, though follow-up studies were never recorded.

A 1955 report by Graham concludes that there was no significant difference between the bottom life forms of fished and unfished areas, but this study, too, was limited by the scientific techniques available. Graham had concluded that tickler chains used in the plaice fishery of the North Sea had not seriously damaged fish food species since catch data compared favourably with other areas not fished with heavy gear. It was noted, however, that heart urchins and paddler crabs had been damaged and colonies of the reef-builder worm Sabellaria had no doubt been destroyed. The destruction of the Sabellaria colonies was lightly regarded because they were considered rare and of little significance in the deeper waters being studied.

The Gear and Behaviour Committee meeting in 1973 benefited from these two studies as well as another series of eight conducted by its membership from 1970 to 1972. A number of conclusions were reached, forming the basis for continuing research, which has continued to the present day. The committee concluded that otter trawls, beam trawls, and rakes or dredges have similar effects on the sea bottom. All have at least short-term negative effects on life forms as attached species are uprooted from their place or broken, while shellfish frequenting the bottom are often crushed or dismembered by the action of the fishing gear. Shrimp beam trawls were seen to be the least destructive; otter trawls and beam trawls using tickler chains were more destructive; and rakes and dredges were ranked the most damaging. The addition of tickler chains and chained groundropes were observed to add to the damage as would a chain mat used to protect the netting of the trawl when fishing rough or stony ground.

The same studies also showed that the effect on the sea bed of passing trawls was dependent upon the nature of the bottom. Hard, sandy bottom stood up well to the gear, while muddy sand showed the effects more readily. In the latter case, a ridge of soil sometimes pushed up as the gear ploughed along. The action of the tides' influenced the lasting impression made by the gear. In areas of heavy tide, the marks are erased more quickly, while in quieter areas, the marks remained for hours or days, depending upon the strength of the current and the bottom type.

A controversial effect of beam trawling that will later be explored in greater detail, is the creation of sand and silt clouds which, when they settle, often smother bottom life forms. Coarser sand was seen to settle more quickly and nearer its original location than muddy sand, which remains in suspension longer and is transported sometimes a considerable distance by current and tide. Successive chains of the gear scrape more and more of the bottom, adding to the cloud and the subsequent damage. In addition, studies reaffirmed the concern over groundropes and chains dislodging boulders from the floor, thus increasing the magnitude of change.

The most extensive study presented to the Gear and Behaviour Committee was made by De Groot and Apeldoorn (1971) of the Netherlands Institute for Fishery Investigations. They reported the Dutch fishery had the most to lose from these investigations into the impact of fishing gear on benthos, since it had been most aggressive in its use of the heavier gear and expansion of its fishing grounds to the French Atlantic coast. The De Groot and Apeldoorn study carefully documented the effects of a beam trawl in relation to the number of tickler chains used to fish for sole and the beam trawl's impact on life forms on the ocean floor. The impact on some life forms was devastating, while others were apparently unaffected by the fishing. The catch rate of some bottom life forms increased with the number of tickler chains to more than three times that with no ticklers at all.

The ICES publication of early findings on the damage caused by bottom trawling and dredging established a benchmark for further investigation.

Continued improvements in sampling techniques and measurement of impact assisted researchers in their compilation of new data. Researchers were then able to expose these early observations and conclusions to closer scrutiny and observation.

A Canadian study which focused on impacts of dredges and trawls on a scallop ground in the Gulf of St. Lawrence was also included in the 1973 submissions to the ICES Gear and Behaviour Committee. The study, headed by Caddy at the Biological Station in St. Andrews, New Brunswick, showed that the dredge cut grooves several centimetres into the gravel bottom, leaving exposed the sand component below. The results indicated that nearly 20 per cent of the scallops left behind had been fatally damaged. It was also noted that damaged scallops would add to the kill as they would be weakened, thus vulnerable to attack by predators. A previous study by Caddy had noted that scavenging fish and predatory invertebrates were attracted to the tracks left by trawls. The 1973 study confirmed that finding. Three species including winter flounder, sculpins, and rock crab feasted on the damaged scallops and other life forms.

Physical damage to the bottom was also noted in the 1973 study. The sediment type and the lack of strong tides in Chaleur Bay, where the tests were conducted, explained why the sea bed was slow to recover. Caddy concluded that repeated dragging of the bottom would jeopardize sea life because a cumulative effect would result in habitat destruction. Particular note was made of the tendency of scallop dredges to lift large boulders from their fixed positions and overturn them, which led to destruction of life forms growing on their surfaces. Caddy concluded that levelling of the bottom by dragging destroys rugosity and the micro environment of species, which rely on stable surroundings. While free-swimming species can search out an alternate location, attached life forms are not so adaptable.

The commercial scallop grounds off southern Nova Scotia and on Georges Bank are deep-water fisheries of greater than 30 fathoms. These grounds of high species diversity on Georges Bank and in the Bay of Fundy are much more dependent upon the attachment of life forms to the surface of boulders and rocks on the seafloor. Studies of similar deep-water conditions in the English Channel, conducted by Holme and published in 1993, have shown that scallop dragging reduced species abundance as forms attached to disturbed boulders failed to survive the uprooting.

Changes in the species composition along the seafloor were given particular attention in a 1982 report by Reise, who studied the Wadden Sea, situated off the northwest coast of Germany. This area had been studied since 1869, so it provided an excellent opportunity to observe changes in species abundance. The Reise study did reveal that some species do, in fact, thrive when their homes are turned upside down by oncoming trawls. Polychaetes, or bristle worms as they are commonly called, were shown to be thriving in the heavily fished Wadden Sea. More than half of the species which were considered rare or missing from the early records but present in 1982 were polychaetes. The increased population of bristle worms met little resistance from beam trawlers in the area as these are a prime food source for the targeted sole and other flatfish.

The same study also showed that a total of twenty-eight species of polychaetes were in decline by 1982. Of those, eight were associated with oyster beds that were fished out by 1925. The most notable loss, however, was eight species that had flourished in conjunction with the tube-building polychaete worm, Sabellaria. The Sabellaria reefs were intentionally destroyed by shrimp fishermen, who found that these were posing an obstacle to their efforts by entangling and ripping their nets. Heavier gear was then employed in a deliberate effort to level the reefs. In the process, the ecosystem—built on the back of the tube worm—was destroyed.

Similar stories of disregard for the marine environment are told today by fishermen of Atlantic Canada's coast, who point out that high-powered draggers and trawlers have been known to remove their nets and load up their footgear in an effort to level the rough terrain on the productive fishing grounds. The effect has been devastating, with peaks and ridges flattened and troughs and depressions filled with boulders, sediment, and bottom life forms. Hook and line fishermen, who use arguably less obtrusive methods, are especially critical. Canadian authorities still have not recognized this deliberate and calculated practice, although it has been brought to their attention during public meetings on the future of the groundfishery.

One of the life forms most vulnerable to the action of trawls is the treelike coral. Hook and line fishermen in Atlantic Canada and elsewhere have learned through years of experience to associate the presence of treelike corals with habitat preferred by groundfish such as cod, haddock, and pollock. These "trees," formed by sedentary marine animals, including anthozoans and bryozoans, help to establish shelter for smaller fish and juvenile groundfish, among other elements in the food chain. These trees are thus a vital link in the maintenance of a suitable micro habitat for groundfish. Studies like Saxon's in 1980 have shown that even in warmer waters, these treelike corals grow very slowly—some at a rate of only 6 mm a year—so, heavy trawling over these grounds could be expected to wipe out these coral beds, further reducing the preferred range of groundfish stocks.

The diminution of a prominent species or of a bottom colony of an area from bottom trawling can lead to an opportunity for another life form to establish itself. Interestingly, scientific advice provided as recently as September 1994 to the Fisheries Resource Conservation Council (the body charged with recommending quotas for Atlantic groundfish) indicates that flatfish stocks off eastern and southern Nova Scotia are in great shape. At the same time, indications are that there is increased fishing pressure on these stocks due to the decline in other groundfish resources. With the health of two of the four stocks at an all-time high, perhaps this bottom habitat has undergone fundamental change with the shift favouring polychaetes—the favourite food of flatfish. This could be the result of the cumulative impact of years of habitat altering activity.

Scientists have come a long way from the Graham study of 1955, which suggested that trawling had little impact on the ocean bottom. Techniques of measurement have improved immensely. While most of the studies have been conducted in the northern hemisphere, scientists in Australia, New Zealand, Indonesia. Thailand, and India, have been actively responding to the concerns expressed by critics of bottom trawling with a series of inquiries. For example, the Indonesian government, concerned that overfishing was depleting the stock and threatening the survival of coastal communities, imposed a ban on the trawling of prawns (shrimplike shellfish). A study by Chong in 1987 indicated that the trawling moratorium had been effective. Prawn stocks recovered and local fishermen, using the traditional labour-intensive method of netting the fish, found they were again able to record good catches and with less effort than when they were in competition with trawlers and purse seiners.

The recent failure of the groundfishery has served to step up Canadian research on the effects of trawls and dredges and to bring into focus the question of which technology and harvesting methods are best suited to the establishment and maintenance of a sustainable groundfishery off Atlantic Canada's coast. A 1991 review by Jenner et al. of a series of pre-existing sonar records and videotapes of the ocean floor of the Scotia-Fundy region was carried out for the Industry, Services and Native Fisheries Branch of the DFO on different kinds of bottom dragging, including groundfish trawls, scallop rakes, and hydraulic clam dredges. The report showed that groundfish trawls were responsible for the majority of the damage to the seafloor because of their popularity.

The videos reviewed by the research team showed that the doors of the groundfish trawls posed a greater threat to the sea bed than the footgear. The doors were seen to plough through the top sediment layers of the floor and occasionally "fly" as they struck obstacles such as boulders. The team concluded that the crashing of the doors undoubtedly crushed animals where they touched down. The doors created a cloud of sediment sometimes more than 50 cm (20 in.) in height as they were pulled along with each door dragging 10 cm to 30 cm (4 in. to 12 in.) into the sediment. Attached plants or slow-moving animals were apparently buried or crushed by the oncoming doors, which vary in weight from 450 kg to 2,000 kg (990 lbs to 4,400 lbs).

The footrope does its share of damage to the bottom. Located just ahead of the lower jaw of the trawl net, the footrope is equipped with ten to fourteen bobbins, each 12 kg to 32 kg (26 lbs to 70 lbs), spaced along its 24 m to 32 m

(26 ft. to 35 ft.) length. The films showed that the bobbins rolled and were dragged through the upper sediment layers of the ocean floor upsetting sand dollars, crab, and clams and tearing away at sea anemones and sea cucumbers. Jenner et al. found that "small animals such as sea stars and bivalves were picked up and churned with the suspended sediment; lawns of polychaete tubes were ploughed up by the bobbins of the footrope. These polychaete tubes appeared very fragile and were possibly destroyed."

Jenner et al. also showed the action of scallop rakes is a serious threat to benthic communities on which groundfish are dependent. These rakes are designed to dig deeply into the sediment to root out scallops, but this action is indiscriminate, tearing away at plant and animal life, both on the surface of the floor and beneath it. The Jenner et al. review states, "In one video a scallop rake being dragged along a level sea bed inhabited by large patches of sea cucumbers uprooted the sea cucumbers from the sea bed by their branchial trees and tumbled them back into the collecting bag. From the video it appeared that the branchial trees were torn from the animals as only the bodies were seen tumbling back over the rake." Other uprooted species included skates, flounders, sea stars, crabs, anemones, and various large invertebrates. These were driven back into the collection bag where the fragile individuals were crushed by rocks, boulders, and shellfish. Although the highly-prized lobster tried to avoid the rake, not all were successful; some were damaged by the passing gear, while others were catapulted back into the collecting bags.

The hydraulic clam dredge is used by the large offshore companies such as National Sea and Clearwater Fine Foods to fish for quahogs and the Stimpson surf clam on Banquereau Bank, for example, located about 80 km (50 mi.) off the coast of Cape Breton. In their 1991 study, Jenner et al. showed this to be the most harmful of the three fishing techniques which scrape and claw the ocean floor. The hydraulic jets reach furthest into the floor, crushing both target and non-target species by its very force. Often both halves of the shells were crushed and those left behind quickly became a meal for a number of scavenger species. Banquereau Bank is also home to commercial groundfish stocks, including cod, haddock, redfish, halibut, and other flatfish, so damage inflicted here and elsewhere by the hydraulic clam dredge has an impact on groundfish habitat. The only mitigating factor with the hydraulic clam dredge fishery is that each company presently has only a small fleet. This could change should government, looking to direct more effort away from the groundfishery and under pressure from the offshore, agree to license more vessels.

Draggers and trawlers have been a favourite target of criticism over the years, even when the only evidence against them was circumstantial and the suspicion was solely the "gut feeling" experienced by concerned fishermen. A dramatic decline in stocks was witnessed whenever and wherever draggers and trawlers entered the fishery. Old-timers complained that too much gear was going into the water and stocks were being attacked at spawning time when they were most vulnerable. They were adamant that fish should be allowed to spawn in peace to better ensure the future of the stock.

Beyond the "gut feelings," there are facts to support growing concern for fish stocks. The harvesting of stocks by mobile gears such as draggers and trawlers is indiscriminate; the methods do not allow for selection of size or species. Any unwanted resource that is caught is simply discarded. Furthermore, evidence continues to build in support of arguments that these harvesting methods are destructive to the ocean bottom and the ecosystem so vital to the commercial groundfish stocks. Harvesting plans which utilize powerful mobile gears are short term and by their very nature unsustainable.

This harsh truth is not lost on young people who live in coastal communities that have been hit by the groundfish crisis. A survey carried out by Barrington Municipal High School students and published in *The Guardian Newspaper* on May 15, 1990, indicated that Barrington and area communities believed draggers were the major cause of the decline in groundfish stocks. In response to a question on the major cause of the decline, 52 per cent blamed draggers; 22 per cent blamed foreign overfishing; 19 per cent blamed overfishing by the domestic fleet; 4 per cent said there were too many licensed fishermen; and 3 per cent identified gillnets as the main culprit. Asked if all fishermen should go back to a hook and line fishery to conserve the groundfishery, 78 per cent agreed.

A conversion from otter trawls to hook and line gear was suggested in Rapporteur's Report, a publication of a groundfish seminar on the Gulf of St. Lawrence fishery held in 1980. Here the Maritime Fishermen's Union (MFU) called for vessels equipped with otter trawls to be converted to fixed gear hook and line. Guy Cormier, President of MFU, pointed out that in the Gulf of St. Lawrence the mobile gear sector had increased in tonnage by 300 per cent between 1956 and 1969. Further, small draggers of less than 15 m (50 ft.) appeared in the mid-1960s, and the use of gillnets also became popular during that same period. As a result, the average size of cod dropped from 4.5 kg (10 lbs) in 1945 to 2 kg (4 lbs) in 1963 and continued to slide with increased pressure from the dragger fleet. Cormier called on the DFO to consider the socio-economic needs of coastal communities along the Gulf of St. Lawrence. These communities were already feeling the effects of fishery downturn as the inshore fixed gear sector experienced poor catches and a shrinking quota. The DFO had, with the extension of Canadian jurisdiction to 200 miles in 1977, granted the dragger and trawler fleets a greater share of the total allowable catch (TAC) available to the domestic fleet. In 1980 the MFU called on the DFO to be cautious in granting quotas to "favour a substantial increase in the biomass"; to show preference for fisheries, such as those using hook and line, which are more selective in their catches; to

Fishing pressure has not only decimated stock numbers but has resulted in a severe reduction in the average size of groundfish. This beauty, dressing out at 34 kg (75 lbs), was caught on handline gear by Max Nickerson of Port La Tour, N.S., in 1976. His son Bruce is pictured holding the cod, which was hooked only 5 km (3 mi.) from shore.

grant a larger share of groundfish allocations to the inshore; and to encourage the catch of larger fish in order to increase the landed value and to regain market position lost when stocks were fished down to the point where only smaller fish were available.

Two recent studies have scientifically evaluated the longline and trawling methods; their findings endorse the position taken by the MFU. A 1990 study by Bjordal and Laevaster of the effects of trawling and longlining on the health of Pacific cod stocks—a potential model for the Atlantic groundfishery—showed that hook and line fishing was no threat, and longlining might continue even after the TAC for trawlers had been reached. Longline fishing is more selective, taking the larger fish that are more likely to feed on smaller cod. The study also suggests that "if recruitment [new fish achieving legal catch size] to the exploitable population is largely influenced by predation on juveniles, then longline fishing may be more beneficial to recruitment." The study points out that when stocks are healthy, the choice of which method to employ is not nearly as significant or critical as when stocks are in serious decline. Managers must act quickly to limit trawler capacity and encourage a shift to the less damaging fixed gear methods.

A very similar finding was produced by a Canadian team of researchers studying the present domestic fleet working the Scotian Shelf. O'Boyle et al. concluded in 1991 that a longliner fleet could fish without regulation because it would not be economically feasible for it to overfish the stock. For longlining to be successful, the fish must be hungry and attracted to the bait. With dragging, the fish could be "sick" or ready to spawn, but they are still very susceptible to the herding by the trawl.

While a passive longline fishery is more readily self-regulated, the dragger fishery is more easily given to overexploitation because of its sophisticated fishing power. Unless strict management measures are put into effect to set limits, trawling vessels are capable of chasing stocks down to the last fish. O'Boyle et al. raised the question: "Why spend taxpayers' dollars in a desperate effort to control an uncontrollable machine and prop up that same machine by granting it half of the groundfish allocations available, when a more efficient, less demanding sector would produce the same end product with much less costly supervision and management."

Further, the study showed that the maximum profit of the Scotian Shelf dragger fleet would be realized only when its capacity was cut in half from the 379 licences available in 1985 to a limit of 150 licences. O'Boyle et al. showed the coexistence of draggers and longliners is virtually impossible. Draggers and trawlers select for younger stock and intercept the cod and other groundfish before they can become large enough for the longline fishery. It was conceded that if draggers did not intercept the stock before it matured, then "longliners would out-compete the trawlers owing to a better revenue-cost picture." Longliners generally take fish in the range of 9 kg to 14 kg (20 lbs to 30 lbs), while draggers target fish of 1 kg to 3 kg (3 lbs to 7 lbs). Trawlers would have to fish three to ten times the number of individuals that longliners would in order to land the same total weight. The case against the gluttony of mobile gear grows when the tremendous discard rates of draggers and trawlers, compared to the negligible rate shown by longliners, are factored in.

The longline fishery is by far the most efficient of the fisheries. Longliners require a smaller capital investment, are cheaper to operate, employ larger crews, and produce a higher value raw product. The present profile of the east coast fishery shows the dependence of coastal communities upon the fixed gear hook and line fishery: the fixed gear fleet has 2,700 vessels, many of which are handliners, while the dragger fleet has only about 150 vessels. However, both have approximately the same amount of quota.

There are at least two sides to every argument, and not everyone agrees that an unregulated hook and line fishery is completely benign. Dragger fishermen are a vocal example. In defence of the sector's record, the executive director of the Nova Scotia Dragger Fishermen's Association, Brian Giroux, remarked that the longline fleet has also been shown to have the ability to catch undersized fish, according to an article in *The Chronicle-Herald* of July 20, 1993. He cited a 1992 DFO observer program report that indicated regulation square mesh used by trawlers and draggers was very effective in selecting larger fish and ranked them ahead of other gear types. According to Giroux, the DFO report showed that the small-hook longline catches more small fish than the square mesh trawl. Consequently, the mobile gear fleet in eastern Nova Scotia was allowed to proceed in 1993 using square mesh. The DFO required the longline fleet to submit a fishing plan to the minister, which indicated its intent to increase their hook size, from a No. 10 to a No. 14 hook, to reduce the take of undersized fish. The No. 12 minimum hook size was made mandatory for the 1995 fishing season as a result of that plan.

The proposed conversion from mobile gear to longline gear is not an ideal alternative to a dragger fleet. Questions have been raised as to how practical it would be to have all dragger fishermen convert their operation to longline. Many in the longline fishery have also intensified their efforts in recent years and to that end have increased the amount of gear they fish. Using automatic baiters and the most modern of hauling gear, each boat is capable of setting miles and miles of longline. Boats typically set anywhere from a few thousand to forty thousand hooks. It has become increasingly more difficult to find room on the fishing grounds for this gear and the conversion of dragger effort to longline would further aggravate this situation. Still, a struggle for space on the ocean may be a small price to pay should restoration of fish stocks be the reward.

While the age-old controversy continues over the merits of hook and line fishing versus trawling, the scientific community has given more attention to these issues and has an important role to play in the debate. Over the past twenty years, a worldwide research base has been established and measuring techniques have been improved. Researchers are able to take more accurate measurement of the effects of fishing methods and gears on target and non-target species and on their habitats. Evidence implicates the fishing methods which employ draggers, trawlers, and dredges. These have been shown to cause both short-term and long-term damage to the seafloor and have, therefore, contributed to the decline of groundfish stocks. Modern research has cast environmental factors in their proper place as only one of a number of influences on the health of groundfish stocks.

Together, the studies make a sound argument for the irresponsibility of fisheries managers and harvesters in the failure of stocks. The shift in research toward fishing methods reveals that management measures of the DFO have been inadequate in their scope and lacking in their enforcement. Research into fishing methods suggests the present poor health of stocks could have been prevented if fisheries managers had effectively exercised their responsibilities more prudently and if fishermen had shown due respect for the natural resource. Excessive fishing pressure and destruction of groundfish habitats are the primary reasons for the present crisis. Optimism and necessity beg the question, what can be done to encourage a return of stock health so that the groundfishery may resume its place in the economy of Atlantic Canada?

Chapter Five

Players for Profit in the Atlantic Groundfishery: Offshore Giants Cast Long Shadows

The commercial groundfishery off Canada's East coast, romanticized by story and song, has always tested those who attempt to eke out a living from the salty billows of the northwest Atlantic. Though few have become wealthy as a result of plying their trade, the groundfishery has remained a vital part of the economy of the region. The industry was slow to develop and resistant to change but persisted under easy sail until World War II. To that juncture, it was characterized as a community-based, low-tech, and small boat inshore operation, struggling against the harsh elements and unpredictable cycles that included low product prices and short supply. Hard work and a commitment to the community's sense of fair play had, for generations, governed the relationship of fishermen with their peers and with the resource. For many in this predominantly small boat inshore sector, the fishery provided a bare subsistence. Prior to World War II, the industry had not attracted the large capital investment that had marked patterns of commercial development in the fisheries of other parts of the world.

The mid-1920s had been a period of very low fish prices in Atlantic Canada. It was a time of transition from the production of dried and salted fish to fresh and frozen product and a time when conditions were generally very unstable in the groundfishery. In 1927 the Royal Commission Investigating the Fisheries of the Maritime Provinces and the Magdalen Islands was established by the federal government to conduct a full-scale inquiry into the problems of the fishing industry and to recommend changes and new policies. The inquiry had been demanded by the industry-dominated inshore small boat sector out of concern that price and supply conditions would worsen should trawler technology be allowed to freewheel. The federal government recognized this concern and moved to strictly control the use of trawlers. With a subsequent ban on the introduction of new trawlers, Canadian operations fishing from the Atlantic provinces were frozen at six vessels, all of them fishing from ports in Nova Scotia. The number of trawlers fishing Atlantic Canada's waters did not increase significantly over the next two decades. In a review of the Nova Scotia fisheries by Watt, the fishing fleet of Nova Scotia included 6 trawlers, 5 draggers, 56 other vessels over 20 tonnes (gross registered tonnage), 5,738 boats under 20 tonnes, and 4,032 small boats by 1946. As the advanced fishing nations of the world increased their catching ability and capacity, the Atlantic region remained a small boat fishery with only marginal growth in the yearly production of groundfish until after World War II.

A sense of optimism swept the Canadian economy with the end of the war. Private capital was attracted to the east coast fishery as economic conditions improved; the demand for fresh and frozen fish products increased; and the industry experienced a period of relative stability. This new optimism provided the impetus for change at the boardroom level as well. The trend was toward merger and amalgamation, and the result was the formation in 1945 of National Sea Products (NSP) in Lunenburg, Nova Scotia. This newly created company drew on the strengths of a number of small independent companies involved in the Atlantic groundfishery. The merger marked the beginning of greater concentration of capital in the industry. With ownership of a trawler fleet for catching fish and processing plants for handling fish, vertical integration, or company control of the stages of production, began to characterize the east coast groundfish industry. The ban on development of a Canadian trawler fleet was lifted in 1949, and a program of government incentives was designed to assist in the modernization of the domestic fishery—the offshore sector was on its way to rivalling the inshore. By 1962, the offshore trawler fleet included thirty-seven vessels. For some, this was the anticipated sign of progress and of industry maturity—east coast Canada was equipping itself to compete with the great fishing nations of the world.

In contrast to the breakneck pace of transformation in the offshore sector, the inshore sector, in many quarters, continues even today in much the same mode as it has for generations. Independent fishermen most often own their own vessels and hire a small crew of one or two deckhands. Sometimes they operate on a share basis with other fishermen, which means the captain uses the money from sales, called the boat share, to first pay expenses, and whatever is left is divided among the captain and the crew. All are driven by the need to make payments on their boats and gear and to provide for their families. According to the Cashin Report of 1993, only in the best years is any money set aside. Inshore fishermen have consistently registered incomes significantly below the regional and national averages.

The scale and type of operation carried out by this community-based, labour-intensive sector are in response to the conditions and opportunities experienced on the inshore grounds. The vessels used by inshore fishermen typically measure less than 14 m (45 ft.) in length and are used to work the near-shore grounds for groundfish and often lobster and other species in season. They use a variety of fishing gears, including hook and line, gillnets, and traps. The boats and crew normally return to port each day, selling their catch to a local processor at the going price. Few options are open to independent fishermen. Even in good years, few are fortunate enough to be gainfully employed year-round as bad weather and ice conditions most often determine the length of the fishing season. With the crunch in the groundfishery, more and more dependence has been placed upon government programs of assistance, such as Unemployment Insurance.

The inshore has seen transformation and modernization, however. Fishermen have shown a remarkable willingness to abandon traditional ways when change offers an advantage in a competitive fishery. Many in the inshore sector upgraded to larger boats and more sophisticated gear in an attempt to better equip themselves for the race to the fishing grounds and to the harvest. This receptiveness to change found salient expression in the ready adoption of the dragger method. Once inshore fishermen were convinced that the new technology could net them a good return—sufficient to offset the initial investment—the shift was on.

A new sector was created, nearly overnight, with the licensing of stern draggers under 20 m (65 ft.) in length to go with the under 14 m (45 ft.) class. With more and more powerful boats taking to the water, it was not long before the unthinkable happened—the yearly dragger quota was fished up before the end of the calendar year. Sector quotas had been assigned after the DFO was handed an all-encompassing mandate in 1977, making it responsible for the proper management of all fish resources inside the 200-mile limit. The fishing up of quotas reached crisis proportions for the inshore dragger fleet fishing the Scotian Shelf. A DFO press release titled "Action Plan on Overcapacity" shows that the four hundred active inshore draggers were twice the number needed to catch the yearly quota assigned this sector, while an equal amount of inactive capacity remained ashore, though it was legally entitled to engage in harvesting. In 1988 and in 1989, the fishing season was interrupted and later closed as the powerful draggers landed their sector quota in record time.

The same investments that had put the inshore dragger fleet at the head of its class also forced its owners to exert greater efforts to pursue a shrinking resource. The intense competition forced everyone to abandon caution in the pursuit of profitable trips. Profits realized were quickly ploughed back into the operation as the latest fishfinding or navigational equipment was purchased to give the operation any added advantage. The race to keep abreast of technology proved to be endless as each advance in catching power was superseded by others. This reinvestment continued, despite the fact that the fleet was more than capable of taking the entire assigned quota before midseason. Some enterprises found it increasingly difficult to land the necessary volume of catch needed to meet expenses. The pressure was on, and the stocks took a pounding.

The offshore sector continued to grow through the postwar years to a place of dominance. The "Annual Statistical Review of Canadian Fisheries Marketing Services Branch" reported that from 1958 to 1968 the capital stock in the Atlantic coast fisheries increased nearly fivefold, valued at $33,306,000 in 1958 and at $168,097,000 in 1968. Much of this growth involved the production of new offshore trawlers whose construction benefited from government subsidies and low interest loans. In Nova Scotia in 1970, three 46 m to 47 m (150 ft. to 155 ft.) stern trawlers were built by the offshore to which a 48.1 per cent federal subsidy was applied with an additional 25 per cent paid through government-guaranteed low-interest loans. In the same year, subsidy was provided for only four small boats—four 13 m (44 ft.) longliners—which were subsidized at a rate of 22.8 per cent of their cost. A "Fisheries Industry Profile and Impact Study" shows that between 1967 and 1970, the federal government spent approximately $526,000 subsidizing Nova Scotia boats longer than 26 m (85 ft.) in length, while it extended only $12,000 in subsidies for the construction of inshore longliners. In addition, the federal government poured large sums of money into the fish processing sector through the postwar years; a large portion of this investment went to the expanding offshore sector. According to Barrett and Davis who wrote "Floundering in Troubled Waters," for the period 1974 to 1982 alone, the federal Department of Regional Economic Expansion (DREE) provided over $64 million in assistance.

As a result of the rapid expansion of the trawler fleet and shore-based processing facilities, the offshore catch and production in Atlantic Canada increased dramatically. In 1966 the offshore catch exceeded the total inshore take. It was a sign, heralded by some, that Canada's fishing industry had finally matured to surpass the labour-intensive inshore sector—capital-intensive enterprise had gained its proper place. Many others, in hindsight, would suggest that government had built a house of cards by encouraging overcapitalization that could not be sustained by the exploitative harvesting rates or the uncertain marketplace.

While government's preferential treatment of the offshore sector was unmistakeable, one company—National Sea Products—received the lion's share of the support. The company underwent two significant periods of expansion, one in the early 1960s, and another in the early 1970s. Though domestic and company catches fell between 1959 and 1961, groundfish prices increased markedly, inspiring National Sea to undertake a major expansion. With the aid of $3.5 million in government assistance in 1964, National Sea built the largest fish processing plant in North America at Lunenburg, Nova Scotia. Following construction at Lunenburg, company plants in Lockeport,

Louisbourg, and Halifax, Nova Scotia; Shippagan, New Brunswick; and Rockland, Maine underwent expansion and modernization. After upgrading its catching capacity through the 1960s, landings declined through the early 1970s. The National Sea response was to increase trawler catching capacity again in an effort to meet market obligations. In 1971 the company expanded to Newfoundland, positioning itself to exploit the northern cod stocks. National Sea was already experiencing warning signs that should have indicated it had overcapitalized. Failing the continued discovery of new sources of raw product, the company was certain to be faced with shortages of supply.

The answer appeared to be the extension of Canada's territorial waters and the expulsion of the foreign fleets. The volume of catch taken by the distant water fleets would then be available to the domestic fleet, hypothetically guaranteeing prosperity for both the inshore and offshore sectors. With the 1977 extension of Canadian jurisdiction, the offshore sector, which had already enjoyed a preferred status in Ottawa's bid to regain control of the Atlantic coast fishery, continued to be the focus of attention. The Canadian government had publicly announced its dedication to restoring groundfish stock health and industry stability. The stage was being set for yet another round of industry expansion, though companies like National Sea were already in debt from having to fund their portion of the earlier growth.

Corporate interests were soon jockeying for position and control in National Sea. Before the end of 1977, a takeover of the company had been accomplished. A rival, but much smaller company, H. B. Nickersons Ltd., acquired outright control of the industry leader. With the assistance of hundreds of millions of dollars from the Bank of Nova Scotia, National Sea then set out to enlarge its catching and harvesting capability. According to Cameron in "Net Losses—The Sorry State of our Atlantic Fishery," considerable investment was made in trawlers: $2 million in 1978, $12.7 million in 1979, and $18.3 million in 1980. At the same time, investments in land, buildings, and machinery for fish processing plants went from $4 million in 1977, to $21 million in 1979, and $10.6 million in 1980. The rate of investment exceeded the profit return and NSP's yearly debt leapt from zero in 1977 to $17 million in 1980. The Kirby Report, titled "Navigating Troubled Waters: A New Policy for the Atlantic Fisheries," showed that yearly net income dropped from a $10.6 million profit in 1978 to a $1 million loss in 1980.

The reasons for the heavy losses were obvious. Under new management and in its rush to sweep the ocean for profit, the company had quickly become overcapitalized and inefficient in its operation. The unregulated expansion resulted in facilities that could not be fully used, and the debt burden could not be paid down. The larger trawlers were expensive to operate, so it was necessary to consistently return with good catches in order to justify their operation. A slump in the fish market price and an increase in the Canadian dollar relative to the American dollar made production marginal at best. As a result, NSP found it could not buy its way to prosperity. The unrestrained purchase of harvesting and processing capacity had become a burden, and rather than guarantee prosperity, it had made the company insolvent. What NSP needed was a fresh start and guaranteed access to a significant portion of the groundfish resource.

The crisis in the fishery again demanded a government response. The Task Force on Atlantic Fisheries was commissioned by the federal government in January 1982 to chart a new course. The resulting Kirby Report recommended a restructuring of the processing sector with a centralization of capital in the hands of a more efficient and professionally trained management. The report also urged the federal government to institute a new resource allocation system, which would award quasi-property rights or individualized quotas to enterprises involved in the harvest of the resource. The stage was now set for heavy government intervention.

In order to salvage the offshore as a significant player in the northwest Atlantic groundfishery, the federal government, with the assistance of the provinces, stepped in with a financial bail-out that resulted in the 1983 creation of Fishery Products International (FPI) in Newfoundland and the propping up of National Sea in 1984. The federal government, in particular, appeared unconcerned with the widespread criticism that millions of taxpayers' dollars should not be used to refloat a sinking ship—corporations like NSP that pride themselves on exemplifying the free enterprise system should be allowed to die by that system as well. Ottawa withstood the storm of protest and carried on with its grand design, which ensured that FPI and NSP, the two largest offshore companies to rise from the ashes, were both beneficiaries of the largest bail-out in Atlantic Canadian history.

This act of partiality provoked inshore fishermen. From their perspective, the government had used their tax dollars to prop up inefficient corporations, only to have them compete unfairly with the small boat sector for a limited resource. The tremendous catching power of the offshore and its mandate to fill a company order, called a "shopping list," which specified the species and size of the catch, often led to conflict between the two sectors along the Atlantic coast. Stories abound among inshore fishermen of offshore trawlers towing their gear over traditionally inshore grounds. They point to the evidence left behind by the trawlers, which often includes destroyed longline gear and gillnets. An Environment and the Economy Symposium on February 12, 1994, revealed that hard bottom—generally the most productive groundfish habitat on the continental shelf and had proven too damaging to trawl gear—was made accessible by the use of adaptive gear towed by vessels with more powerful engines. The traditional grounds favoured by the inshore and midshore hook and line fishery had shrunk appreciably due to encroachment from large offshore trawlers that could then access waters closer to the shoreline.

More importantly, any chance for groundfish stock recovery from decades of foreign overfishing was thwarted by the increase in catching and fishing power of the offshore trawlers. The recovery during the years 1978 to 1982, when Canadian catches of all groundfish species increased from 535,000 tonnes to 775,000 tonnes, were short-lived; the downward slide again gathered momentum as catches fell to 418,000 tonnes in 1992, the year the harvest throughout much of the region was pre-empted by an imposed moratorium. Any hope of prosperity for the inshore was pirated by the offshore companies in the pursuit of corporate profits. The inshore sector from Labrador to southwestern New Brunswick complained bitterly about the unfair advantage held by the offshore sector and their aggressive approach to the harvest, without effect. Fishermen concluded that their concerns had fallen on deaf ears. After all, with major investments in the offshore companies, it seemed unlikely that government would impose limitations on the trawler fleet in order to restrict its activities on the inshore grounds and ultimately reduce its profit.

These suspicions were confirmed in 1982, when an Enterprise Allocation (EA) program designed to meet the specific needs of the offshore sector was instituted on a trial basis. The EA program gave offshore companies property rights to fixed amounts of fish to enable them to break the cycle of glut and shortage, which had contributed to their insolvency. Members of the inshore sector were outraged, believing the EA program would allow for a greater portion of the harvestable resource to go to the offshore than was justified. Knowing that they could take more than one-half of the total Atlantic groundfish catch under the new arrangement, the offshore companies were afforded the luxury of being able to organize and manage their harvesting efforts in such a way that expenses were reduced and the greatest monetary return for the catch was realized. This meant that trawlers could pick the best weather conditions to fish, and they could land product at times which yielded the best market prices and were the least inconvenient for managers charged with organizing the processing function. Enterprise Allocations, or individual quotas by company, were later granted to the midshore groundfish fleet in 1988.

The original split of groundfish quotas between the inshore and offshore was to be based on the historical share of catches recorded by the two sectors. This split was not resolved until after the restructuring of FPI and NSP was complete in 1984. The Cashin Report revealed the offshore share of the total Atlantic groundfish catch had fallen from a healthy 350,000 tonnes in 1982 to 309,000 tonnes in 1983, and 308,000 tonnes in 1984. The same study showed that the inshore catch by comparison reached 470,000 tonnes in 1982, 456,000 tonnes in 1983, and slipped to 427,000 tonnes in 1984 under the quota sharing arrangement. Yet, the first year the split was fully implemented under the 1985 Atlantic Groundfish Management Plan, the inshore was granted a quota of 476,000 tonnes while the offshore was handed an excessive 485,000 tonnes. It was clear to those in the inshore sector that the offshore was receiving preferred treatment.

The Enterprise Allocation system proved to be a godsend for the offshore companies, enabling a reversal of their financial prospects. National Sea and Fishery Products International went from significant losses in 1984 and 1985 to remarkable profits. According to the Cashin Report, NSP registered profits of $10 million in 1986, $36 million in 1987, and $25 million in 1988. FPI fared even better, recording profits of $47 million in 1986 and $58 million in 1987. In an article titled "Net Losses—The Sorry State of our Atlantic Fishery," Cameron noted that FPI had done so well under the Enterprise Allocation program it was able to buy back government shares that had been used in 1983 to finance its launch.

While the EA system, implemented by the federal government, played a significant role in the success of these two offshore giants through the mid-1980s, the fishing industry also benefited from lower fuel prices, a weak Canadian dollar, and the growing strength of fish prices as health conscious consumers discovered the value of fish as a low-fat staple. However, as the success of the domestic offshore sector grew, it had to face stepped-up opposition from the inshore sector which accused trawler fishermen of highgrading (keeping only the fish of highest value and returning lower valued fish to the water); discarding (throwing over the side any landed catch that is undersized or non-target species); and dumping (releasing an entire catch from a dragger or trawler net) under the EA program. Critics were adamant that the graphic description of such destructive fishing practices was sufficient evidence to prompt a public inquiry; they were confident that increased vigilance by the fish police would provide all the courtroom evidence necessary to indict the EA system. No investigation materialized, but members of the inshore sector were convinced the federal government was in cahoots with the two large offshore companies.

Instead of terminating the arrangement, Fisheries Minister Tom Siddon made the EA program a permanent fixture in 1988. In a DFO press release of December 30, 1988, he repeated earlier arguments made by federal authorities, which suggested that rather than have everyone compete for a maximum share of a common quota, the EA system contributed to "a more economically efficient and stable pattern of fishing, more uniform harvesting and processing activity throughout the year and better marketing of the resource." These benefits would, of course, accrue to the offshore companies who could spread their harvesting effort over the year matching best market prices with increased catches, thus garnering the highest return for the volume of product harvested. These benefits had nothing to do with saving the resource and everything to do with economics. The proposed end result of the EA program was a more stable offshore sector with increased earnings for trawlermen and plant workers.

To further encourage the development of the offshore sector and enhance its competitiveness in world markets, the federal government granted a factory freezer trawler licence to NSP of Lunenburg, Nova Scotia, in November 1985. The factory freezer technology was familiar to the northwest Atlantic, having been used for a number of years by the distant water fleets. In 1985 there were a total of 74 foreign-owned factory freezer trawlers licensed to fish inside Canada's 200-mile limit. As early as 1977, a national debate arose as to whether the domestic offshore should be licensed to use the factory freezer trawler (FFT). That same year, the federal government had co-sponsored with NSP an experimental charter of a West German vessel. Under the agreement with the federal government, NSP was to concentrate the trial fishery on non-traditional species. The licence was denied in 1977 and on four other occasions over the next eight years as the federal government changed hands and opposition from fishermen persisted. Inshore fishermen had been opposed to the presence of foreign freezer trawlers, which they claimed were guilty of destructive fishing practices and ruination of stocks. The prospects that the federal government would consider allowing the offshore to shift to such a powerful and pervasive harvesting technology as freezer trawlers was deeply troubling. Newfoundland, concerned about the impact of FFTs on their inshore fisher, which was already in decline from years of foreign and domestic offshore over harvesting, also joined the debate.

With the granting of a licence in 1985, NSP purchased state-of-the-art harvesting technology, paying $8 million for the *Cape North,* an 82 m (269 ft.) West German vessel. Acting Fisheries Minister, Erik Nielsen, set out a number of conditions for the vessel's use, including the stipulation that at least 50 per cent of the catch was to be comprised of previously underutilized EAs. Also, no more than 6,000 tonnes of the company's northern cod allocation was to be harvested by the *Cape North* in any one year. These conditions were included to help quiet the still raging storm of protest from inshore fishermen and the Newfoundland government, who were concerned that the new technology would be unleashed upon migratory northern cod stocks before they left the deeper waters for their yearly migration to the near-shore grounds. The fears of fishermen were not allayed by the government's promise that no additional EAs were to be granted with the freezer trawler licence and that NSP was required to retire equivalent capacity from its existing fleet. They were fearful that government, already committed to preferential treatment of the offshore, would eventually be persuaded to relax these restrictions and the new all-powerful harvesting technology would then be free to ride roughshod over the resource and the inshore fishermen's traditional lifestyle.

A socio-economic impact study was carried out on the operation of the *Cape North* in 1987. The anticipated unemployment was minimal and localized in the area immediately surrounding Lunenburg, Nova Scotia, *Cape North's* home base. Although a potential net loss of 133 jobs had been predicted, a strong market for fish and healthy catches had actually masked the impact of the freezer trawler on shore processing jobs, with an increase of 28 full-time positions at the plant in 1986. This was one of only two years in which the trawler turned a profit as catches rapidly deteriorated and groundfish prices fell from 1988 on. By 1988 the flagship of the fleet was a lame duck, and soon the huge vessel was spending more and more time tied up or in dry-dock, and less time fishing. A company decision was reached. In January 1994, NSP sold the *Cape North,* citing the reasons as closure of the northern cod fishery and reductions in groundfish quotas. The sale was a final chapter in the company's dream of securing a stable domestic supply of groundfish to meet its market commitments. It would have to place more dependence on foreign-owned freezer trawlers fishing the open seas.

While the offshore companies were granted property rights to a large share of the groundfish resource under the EA program, the inshore sector remained a competitive fishery to the end of the 1980s. This meant that each vessel in the inshore was virtually unrestricted by government regulations in the amount of fish it could harvest. Stoppages were imposed only when the quota for a particular management zone or fleet sector was in danger of being exceeded. Government incentives for vessel construction offered after World War II had enticed some, so the inshore grew steadily through the years leading up to extension of Canada's territorial limit.

After 1977 there was a marked increase in the construction of inshore vessels and of draggers in particular. The problems inherent in this build-up became readily apparent through the early 1980s as catches dropped, despite an increase in the effort exerted by the inshore fleet. While the overwhelming majority of groundfish licences and boats in the inshore belonged to the fixed gear sector, the most powerful segment in all parts of the Atlantic region was the mobile gear dragger fleet. According to the Rapporteur's Report for the Gulf Groundfish Seminar held in Memramcook, New Brunswick, in 1980, fixed gear fishermen were soon registering complaints that the dragger fleet had an unfair advantage. A growing chorus suggested that the slide in the east coast fishery began with adoption of modern inshore harvesting methods and, in particular, with the use of bottom destroying draggers. The fixed gear sector suggested this method was adopted simply because it was the easiest way to land large volumes of fish, even though the method showed little consideration for the fragile balance supporting targeted groundfish stocks and their place in the marine ecosystem. There was a growing imbalance between the catching and finding capacity of the inshore fishery and the gradually declining resource of groundfish stocks.

Yet, bottom dragging methods are those of choice for both inshore highliners and the giants of the offshore. Dragger and trawler technologies allow a high-volume harvest while keeping the per unit costs at a minimum, maximizing profits while stocks are in good supply. When stock health fails, however, these harvesting methods quickly lose their competitive edge. The huge investment in vessels, equipment, and onshore processing facilities means that in short supply periods, these enterprises are no longer economically viable. It is then time to divest of vessels and close plants, a drift all too familiar to those following the Atlantic coast industry over the past ten years.

In response to the moderation of traditional harvesting methods, the following sentiment has been expressed time and time again by a number of experienced fishermen and industry participants, who have watched the changing in approach to the fishery, in method and in attitude. Basil Blades of Sable Fish Packers, Cape Sable Island, was quoted in *The Guardian Newspaper* on May 16, 1989, commenting on the state of the fishery: "Fishermen are interested in catching fish the easiest way they can. And the easiest way they can is put on their carpet slippers and go down and turn in their bunk and drag for four hours and then come up and pour them [fish] down the hold round [undressed]. You no longer have fishermen, you have carpet slipper fishermen. Carpet slipper fishermen are the reason stocks have been depleted."

Heavy fishing pressure during the 1980s soon depleted traditional stocks such as cod, haddock, and pollock throughout the Atlantic region. The Scotian Shelf quotas, determined on the basis of stock health, dropped one-third from 1981 to 1988 and continued to fall. Extensive management measures aimed at controlling the inshore fishing effort, including licences, quotas, seasons, and trip limits, failed to compensate for the increased fishing power. The temptation to fish illegally was too great for many members of the inshore as it was for those in the offshore. Illicit fishing practices, such as overfishing (fishing beyond a boat or sector quota), dumping, discarding, and highgrading became common. The rewards for slipping past the authorities with a pirated catch were too great and the punishment was minimal through the early and mid-1980s. Most often the minimum penalty for these offences was issued by the criminal justice system, which seemed to take a romantic view of fishermen as family providers simply trying to make a go of it in an over-regulated industry. It was not until it had become accepted that stocks were at a crisis level that these fishing practices were regarded as theft from another fisherman also trying to provide for his family and boat crew.

Those representing offshore trawlermen were well aware of an aggressive bent of inshore dragger fishermen and were concerned that their continued lobby efforts would win the dragger fleet an even larger share of the shrinking groundfish quota available for the Scotian Shelf waters. This insatiable appetite for more fish quota was seen as a very real threat to trawlermen and plant workers employed by offshore companies as year after year the inshore, led by the dragger lobby, had been able to successfully appeal to the DFO for a transfer of groundfish quota from the offshore allocation. Addressing the

Scotia-Fundy Groundfish Task Force, or Hache Task Force, in August 1989, Kingsley Brown, Executive Director of the Fishing Masters Association, made his views known.

"Anyone who lives in a fishing community knows how much fishermen make. Southwest Nova Scotia dragger fishermen make more money than doctors, lawyers, accountants, judges, and teachers in their region. By demanding more quota from the offshore allowance, they want to take away the jobs of hundreds of trawlermen and thousands of processing workers, mostly women, who are lucky to earn, with overtime, $15,000 per year. There is no moral courage in taking from those who have less. That's how to make a genuine crisis of hardship."

Scotia-Fundy dragger fishermen, despite their relatively short history, have developed a strong sense of identity and single-mindedness and, with that, a justification for their aggressive style of fishing. In addition, this sector has had strong representation and has not hesitated to state its case. While not condoning overfishing or other illegal activities, Clifford Hood, counsel for the Nova Scotia Dragger Fishermen's Association, perhaps best explained, in a "Submission and Response," the operating drive and motivation of fishermen in that sector.

"Often it appears to the members of this Association that they and their colleagues in this area are portrayed as aggressive, rebellious, highly paid or even, as has been recently reported, as 'buccaneers.' This misconception evolves because of a basic misunderstanding by those who are charged with the responsibility of overseeing our activities with the true psychological profile of the western Nova Scotia fisherman. He is, in fact, one of the last of a class of people the anthropologists call the hunter-gatherer. He is a risk taker who not only places his livelihood, his home, and his bank account in jeopardy when he engages in his enterprise, but he places his life at stake when he ventures into the natural element from which he derives his living."

However, in an article in Yarmouth's *Sou'Wester* on September 1, 1989, and in a report by Angel et al., fishermen and fishermen's representatives freely admitted that they underreported and misreported their catch, that they took more than their fair share of the resource. Guestimates place the illegal, unreported catch of groundfish at an additional 50 per cent during the runaway years of 1985 to 1988, when prices were high and quota reductions were prescribed. Catches of haddock and pollock were listed as flounder, or some other species, when quotas of the former were about to expire. Also, catches in a restricted or closed zone were recorded as coming from another to avoid confiscation of the catch. Scientific advice which informed the setting of the next year's quota, was, in effect, based on erroneous input from officially recorded catches provided by fishermen.

The authorities made a concerted effort to control the plunder, with only limited success as the modern day hunter-gatherer also proved he could usually keep one step ahead of the authorities. The many isolated wharves and the cover of darkness allowed much of the catch to go unreported. Trucks equipped with booms for off-loading the boats waited at these wharves, ready to transport the product to the fresh fish markets in New England and Ontario or to nearby processing plants.

Cuts to the inshore quotas prompted calls for transfers of quota from the offshore, which had most often failed to fully utilize their allocations along the Scotian Shelf. Provisions had been made in the annual groundfish management plan for quota transfers as one sector approached the end of its quota. Managers often yielded to the rather excessive pressure applied by the inshore sector and led by the dragger fleet. Occupation of DFO offices, staged demonstrations, and damage to government property were common when the resource was particularly short. One series of especially violent demonstrations in October 1988, in southwestern Nova Scotia, prompted the Minister of Fisheries and Oceans to close the fishery until law and order could be restored.

Some may have chosen to explain this illegal activity, including overfishing and violent demonstrating, as a justified response by desperate men to an oppressive system of fisheries management. Greed and overzealous ambition would better describe the mind-set of fishermen involved in these activities. They had stepped over the line. Veteran fishermen could only shake their heads and point to the dire warnings of earlier generations. Greed, aided by the tools of modern technology, had fuelled the slaughter of groundfish stocks. Even in the good years of the early 1980s, when there were plenty of fish, dumping was occurring, but since the practice at the time did not pose a direct threat to the stocks, fishermen tended to keep their concerns about the waste to themselves. When quotas began to be squeezed, however, fishermen realized that the senseless waste associated with dumping, discarding, and highgrading was going to have an impact on their future.

The wanton disrespect for authority and the disregard for conservation of fish stocks shown by the inshore dragger fleet did draw criticism from other inshore groups. Maritime Fishermen's Union (MFU) Local 9 Executive Director, John Davis, cautioned that excessive overfishing of groundfish stocks, which occurred on Browns Bank in 1988, would be felt for years. The signs from reported catches the following year were frightening, according to Davis in an article from *The Guardian Newspaper* on February 21, 1989. He blamed the lack of fish on overfishing and on the ineffective management by the DFO. He warned that all sectors of the fishery would pay the price for the excessive fishing by the dragger fleet on Browns Bank.

Another voice among the dire forecasts of the future of the inshore groundfishery was Noble Smith, veteran fisherman and president of the Sou'West Nova Longline Association. In an address to the Barrington Municipal Council in October 1989, he predicted the collapse of the inshore groundfishery within two years if the overfishing continued at its present pace. He told Council that it must find alternate sources of employment to offset the coming losses in the fishery. He cited modern technology, which enabled boats to locate and catch enormous quantities of fish, and the dragger fleet which removes about 80 per cent of the total groundfish caught by the domestic fleet for the demise of the industry.

Similar concerns for the future of the inshore groundfish industry were echoed from ports throughout Atlantic Canada. In the Gulf of St. Lawrence groundfish declines were blamed on the intensive fishing and overfishing of the dragger fleet. Stocks were never allowed to recover from the heavy pressure inflicted by the large foreign and domestic trawlers during the 1960s and 1970s. By the late 1980s stocks of codfish, hake, redfish, and flounder were collapsing. Consequently, their biomass plunged to half that recorded in 1984. The end result was an extension of the moratorium on cod fishing in waters off Newfoundland to the Gulf of St. Lawrence on December 20, 1993.

When it became obvious that management measures, such as sector quotas, closed seasons (managed times when grounds were closed), and trip limits (catches were set by the DFO for each vessel going out), could not stem the tide of increasing investments in catching power and the bent of dragger fishermen—especially those in the Scotia-Fundy region—the government turned to individual quotas whereby each boat was assigned a total allowable catch (TAC) for the year. A number of shellfish fleets had operated on individual quotas, so the federal government felt confident that this measure would help slow the race by fishermen to catch the largest possible share of the overall quota. Individual quotas assigned to each inshore dragger were based on that boat's historical catch data, and the total quota for the dragger sector would be subdivided. The individual boat quota (IBQ) system, made permanent in 1991 for all Atlantic coast draggers under 20 m (65 ft.) in length, was ticketed as the solution to the overcapacity and runaway fishing power of the fleet. It was assumed that the system of boat quotas, backed by a more stringent catch monitoring system and penal code, would deter those inclined to cheat. With fishermen able to depend on a secure catch, it was assumed the incentive was in place for each boat owner to tailor capacity and fishing effort to the quotas they controlled. With the introduction of the inshore transferable quota (ITQ) system, profits would be maximized, and successful fishermen would be positioned to buy out those looking to retire from the fishery.

But the new IBQ system, designed primarily to reduce pressure on groundfish, failed miserably. If some cheated before the system, even more cheated after it. With each boat assigned a quota, it became obvious the fish which fetched the highest price should be kept and lower value fish—most often the smaller juveniles and restricted species—should be dumped. Highgrading became rampant throughout the fleet after 1991 because of the boat quotas, the same system that was to bring stability to the groundfish industry, give fish a chance to grow to maturity, and restore stock health. As long as regulations stipulated a fisherman must stop fishing all three major groundfish species once the quota for any one of the three was reached, the incentive to dump perfectly good fish remained.

The IBQ system also spawned a new system of cheating—transhipping. Dragger fishermen found that they could continue fishing beyond their quotas as long as they were able to collaborate with a fixed gear fisherman. The fixed gear fisherman, who still operates under an open and competitive fishery, would, as his part of the deal, receive the transhipment at sea, land the overquota portion, and charge it against the fixed gear sector's total allowable catch. Transhipping was popular and often openly practised with deals boldly carried out on VHF radios. An "under the table" or "over the airwaves" deal would then be reached on splitting the profits from the caper. Some bizarre scenes resulted from these transactions, with hook and line fishermen sometimes off-loading tonnes of fish without evidence of hook lesions a telltale sign that the fish had been covertly captured by use of a dragger's net.

This collusion between dragger and fixed gear fishermen gave rise to a vociferous protest from the fixed gear sector. In September 1991, the fixed gear sector off southwestern Nova Scotia was told it had quota sufficient to last the rest of the year, but only two months later it was closed down. The fixed gear sector blamed the closure on transhipping. The dragger fishermen, for their part, countered with the popular rejoinder, "it takes two to tango," and that some fixed gear fishermen were equally at fault for the sudden closure. Statistics showed that landings of handline fishermen in Digby County, Nova Scotia, nearly doubled with the popularity of transhipping.

Members of the fixed gear sector guilty of trying to make a fast buck via transhipping, were also prone to time-honoured forms of cheating, such as underreporting catches and fishing in closed areas, according to an article in The Guardian Newspaper from October 4, 1989. Underreporting and misreporting were not as prevalent in the fixed gear sector until the quota crunch of the late 1980s. The threat of seizure of an over-the-quota catch provided all the necessary incentive for some members of this sector to also make clandestine arrangements for off-loading their catch under the cover of darkness and beyond the watchful eye of authorities.

Despite evidence that overfishing, highgrading, and dumping are the root of the failure of the Atlantic groundfishery, these activities continue to be practised by the fixed gear sector of southwestern Nova Scotia—the only region not yet closed by moratorium. Horror stories surface to this day which indicate that the fixed gear fleet fishing waters south of Nova Scotia is actively involved in highgrading, discarding twice as many undersized haddock and cod as are landed. At a public meeting held in Clark's Harbour in September 1995, fishermen freely admitted to discarding perfectly good fish at an increasing rate, despite the fragile condition of the stocks, and despite government's efforts to curb destruction of the resource. This insolence exists because the drive to maximum profits fosters little or no regard for long-term stock health. What will it take for fishermen to realize that they are destroying their own future along with their destruction of fragile fish stocks?

Although the inshore sector has been guilty of destructive fishing practices, the scale of this illegal activity pales by comparison with that attributed to the offshore trawler fleet. Angel et al. report that fisheries managers have been well aware that implementation of the EA system has fostered the dumping of massive amounts of fish deemed of lesser value by offshore companies. The horrendous stories, which indicated that big company trawlers were highgrading their catches and dumping as much as three to four times the weight of fish landed, had circulated along the coast for years. The stories were accepted as fact by inshore fishermen. They had fished alongside some of the huge trawlers and had witnessed the mantle of white bellies floating on the tide after the catch had been sorted and the deck cleared, according to an article in *The Chronicle-Herald* from November 26, 1992.

As fish stocks declined, large offshore companies concerned with rationalization began to close fish processing plants. These closures effectively severed the loyalty of many former employees. With the announcement by NSP that it was abandoning the town of Canso by closing its processing facility there, employees and former employees no longer felt obligated to protect the company's interest. According to an article from the *Sou'Wester* on March 1, 1990, Patrick Fougere, the outspoken union representative of trawlermen fishing from the port of Canso, told a press conference held February 17, 1990, that National Sea Products had made it a practice of having its skippers dump perfectly good market-sized fish simply to meet a shopping list provided by the home office. If, for example, they wanted 453,000 kg (100,000 lbs) of cod, 9,060 kg (20,000 lbs) of flounder, and 6,795 kg (15,000 lbs) of haddock, the instruction would be to dump any product beyond that amount. Complaints that NSP trawlermen were fishing from a shopping list had been registered before, but with the entire country focusing on the plight of Canso, the media coverage elevated the issue. The DFO, reluctant at the time to confirm the reports, later acknowledged that they knew the offshore trawlers were undertaking so-called "Sobeys trips." Angel et al. found that some trawler captains, disturbed by the waste caused by fishing from a shopping list, had complained to DFO observers stationed on the vessels, "If I land it my company will suspend me, but if I dump it I am in trouble with DFO."

In fact, NSP had expended considerable effort prior to the media attention in 1990 in an effort to keep a lid on accusations that its trawler captains fished from a shopping list. Appearing before the Hache Task Force in August 1989, Michael O'Connor, Manager of Fleet Services, tried to paint a different picture.

While stating that the company's fishing plans were influenced by many factors, including market requirements, he suggested that catches were no longer highgraded, and fish were no longer dumped or discarded.

"On some trips, National Sea vessels are directed for one species with a 10 per cent by-catch allowed. These would include Northern cod and Gulf redfish. For multi-species trips on the Scotian Shelf, vessels directing for flounder and pollock are permitted a significant amount of codfish to go with their trip. This allows the vessel to maximize their species directed for, when the fishery isn't clean [when cod are found swimming the same waters with other groundfish], along with helping increase the vessel's total catch at the end of the trip.

...There is presently a cod tolerance in place for all vessels when directing for all other species. For instance, all vessel captains have an allotted amount of cod above their allocated trip amount, which they can use when directing for other species. When a vessel overruns its cod trip limit, it may use this reserve to help maximize the species directed for. Contrary to some claims, this minimizes dumping and discarding by our vessels." With this explanation, Michael O'Connor was rejecting accusations made by independent fishermen that NSP continued to require its trawler captains to dump market-sized cod and other groundfish when on a trip designed to land other species.

Henry Demone, President of National Sea, and Murray Coolican, Vice-President of Government Relations, were called by the Nova Scotia House of Assembly, Committee on Resources, to comment on charges that the company encouraged destructive fishing practices. The company executives indicated that they, too, were concerned about the dumping of fish and had instructed trawler fishermen that they would not tolerate destructive fishing practices. Should a captain of one of their vessels find that he is towing in an area that is returning too much of one species, then he is expected to move on to avoid having to dump the restricted species. Demone explained to the House of Assembly, Committee on Resources, on April 24, 1990:

"What I cannot accept from any of our captains or from anybody in industry is that they shoot away [set the trawl] once in this area and continue to fish in this area. I can tell you that this behaviour at National Sea is frankly not tolerated. We have suspended captains for this and if we or DFO find them engaging in this type of behaviour, we will suspend them again. We do not support this, we frown upon it, and we do everything we can to stop it."

Observer reports filed with the DFO during 1989 refute the company claim indicating that "company-imposed trip limits stemming from EAs continue and domestic vessel captains are still 'walking a tight-rope' between their companies and DFO." Angel et al. suggest the reports, dated from 1989 to 1993, showed that though many captains are hesitant to discard "under the nose" of an observer, the crews "freely admit to massive dumping when observers are not aboard."

Former observers who were under contract with the federal government and assigned the duty of monitoring fishing activities from on board ship, point out that almost none of their reports of violations, such as the dumping and catching of undersized fish by offshore companies, were prosecuted. One case did reach the courts in 1993. A Lunenburg County, Nova Scotia, trawlerman, Captain Donald Weagle of West Dublin, was fined $15,000 for fishing with an undersized trawl. The crew had used a second smaller meshed trawl left on board after completion of a previous trip. The smaller meshed trawl would have been legal had the crew been directing for redfish, but when fishing for cod it would have trapped a higher percentage of juvenile or undersized fish—fish which had not yet spawned. His vessel, the *Cape Lantz,* a National Sea trawler, was capable of catching 226,500 kg (1/2 million lbs) of fish per trip but only landed 57,078 kg (126,000 lbs) of cod on that particular voyage in December of 1992. Thus, the potential damage to stocks was much greater than actually realized.

Despite DFO's awareness that NSP trawler captains acted under strict control by the company and that head office innocence in the matter was questionable, the offshore was a direct beneficiary of the incident. It sold Weagle's forfeited catch for only $27,000 (21 cents per pound, when the going market price would have been approximately three times that amount). The sale of a forfeited catch at a reduced price to any interested buyer is consistent with DFO policy. Seldom, if ever, does a second buyer enter in the sale, choosing not to get involved in a matter considered private between the regular buyer and the justice system. Therefore, the fine which is imposed upon a trawler captain is hardly a deterrent for a company which owns and operates a fleet of trawlers and can write off the few occasions, when its operations are slowed by prosecutions, as no more than the price of doing business. NSP, true to the form that President Henry Demone indicated, disowned Captain Weagle and fired the sixty-year-old fisherman. Charges against NSP over the same incident were dropped.

There are reasons for using undersized trawl and landing undersized fish. National Sea has a vested interest in the landing of small fish, as Michael O'Connor, manager of Fleet Services, inadvertently revealed. In explaining why the company was not interested in using larger mesh to allow small, immature fish to escape, *The Chronicle-Herald* of October 24, 1991, reported Mr. O'Connor's remark that all NSP's processing machinery is geared to process fish 25 cm to 35 cm (10 in. to 14 in.) in length. That size is well below the long-established 41 cm (16 in.) minimum length for cod, haddock, and pollock in the Atlantic region. These minimum sizes were implemented following detailed discussions with the fishing industry and applied to mobile gear—draggers and trawlers—and fixed gear—longliners and gillnetters—but did not apply to the cod trap fishery, according to a DFO press release of March 20, 1989. While the smaller fish processed by NSP are destined for value-added entrées—a fresh frozen and battered product that fetches a premium price—the larger fish are culled out for another processing line. O'Connor explained, "If we get a large fish we do what everyone else does, split it and salt it."

As an example of how far NSP went to obtain undersized fish, one Cape Sable Island resident, now deceased, openly related to community members how it was his duty as an employee of the company to knit a net liner once the company trawler left port. The illegal small mesh liner would be used to ensure a hefty catch, including smaller fish, which would be whisked past fishery inspectors into the processing plant. The veteran fisherman explained that the net liner would be destroyed or discarded at sea, thus disposing of the incriminating evidence. In the report by Angel et al., the DFO recently acknowledged the common and illegal use of net liners through the years immediately following extension of Canadian jurisdiction in 1977. They failed to proceed with prosecution, judging that no action was required since stocks appeared to be recovering from previous overfishing. Even today, with observer coverage on less than half the trips taken by the domestic trawler fleet, there is a legitimate and nagging concern that such practices as dumping, discarding, misreporting of species caught and of the area fished, use of net liners, and fishing in closed areas are still carried on by the offshore fleet.

Trawlers, both domestic and foreign, expend time and energy attempting to catch undersized fish because there is a demand and thus a market for them. The Europeans consider juvenile fish a delicacy. Despite North Atlantic Fisheries Organization (NAFO) and European Economic Community (EEC) regulations to the contrary, baby fish were being harvested outside Canada's 200-mile limit on the Grand Banks as recently as 1994. The evidence was landed in St. John's, Newfoundland, in April 1994, after federal fisheries officials boarded the *Kristiana Logos,* a formerly owned and operated Clearwater Fine Foods trawler, which had been sold by the company and re-flagged, or registered, in Panama to elude Canadian quotas and regulations. The vessel had been fishing amongst sixty other foreign vessels just outside Canadian waters and had been under air surveillance by patrol planes dispatched to the area. The crew of the *Kristiana Logos* had been suspected of fishing for groundfish species, including cod, which are off limits to NAFO countries. A lined net used to capture undersized fish was found on deck, and the illegal product, consisting mostly of baby groundfish measuring a minute 10 cm to 15 cm (4 in. to 6 in.) in length, was found below.

Other countries have also been caught in the illicit harvest of undersized fish. In the summer of 1993, *Fishing News International* reported that a British film crew had filmed the land-based portion of Spain's trade in illegal fish. The film crew were acting on allegations from British and Irish fishermen that foreign vessels carrying U.K. registration, manned by Spanish crews, and directed from the Basque region of Spain, were fishing juvenile groundfish off the British coast and landing the small fish at the port of Ondagrrow. (It is also from the Basque region that ships sail to fish off Newfoundland's Grand Banks. In 1995, as a result of Tobin's "Turbot tussle," it was shown that the Spaniards have been involved there in the fishing of juvenile groundfish.) The television crew filmed undersized fish as small as 6 cm (2 in.) in length for sale at the port, and they explored vessels with concealed fish holds designed to thwart fisheries inspectors. The floor carpet in one crew's quarters was pulled back to uncover the entrance to a secret fish hold. The film crew was unable to finish their shoot; they were attacked by angry dock workers. The cameraman was struck on the head with a hook-type tool, while the reporter and producer were roughed up and their equipment was damaged. Fisheries inspectors had made previous attempts to investigate rumours but had been driven back to their hotel by a mob.

A complaint based on the findings of the British film crew was registered by British Fisheries Minister, Michael Jack, with European Community Fisheries Commissioner, Paleokrassas. British fishermen were hopeful that the EC would be successful in its investigation of the matter. A formal investigation could have confirmed that Spanish fishing interests have been involved in the capture of undersized fish off Canada's 200-mile limit. In *The Last Cod-Fish,* Chantraine reveals that foreign trawlers were found to have fished the northwest Atlantic, landing their catches of undersized fish at the French island of St. Pierre, where they were processed at a Spanish-owned facility and sent by ferry and road to St. John's, Newfoundland. The illegal cargo was then flown to Scotland, where the tiny fillets were disguised by a heavy coat of batter and sold to unsuspecting British consumers at a healthy profit.

The Canadian fishing industry, the politicians, and even fisheries managers may prefer to blame much of the present crisis on the illicit activities of the foreign fleets and the actions of high-seas pirates, such as the Basque fleet of Spain. However, the real source of the problem has been identified closer to home. The destructive fishing practices of the domestic inshore and offshore fleet has been largely responsible for the destruction of groundfish stocks. The extent of harm by illegal fishing practices was confirmed in two federal reports, one filed in 1989, the other in 1990. The Hache Task Force Report of 1989 examined the state of the groundfishery in the Scotia-Fundy region and made recommendations on how to better manage the fishery. The Harris Report of 1990 was primarily concerned with the state of the northern cod stock and its future prospects, but it also addressed the threat posed by the continuation of illicit fishing practices.

The Hache Report found that measures taken by fisheries managers to curb growth in fishing capacity in the inshore fleet of Scotia-Fundy had been ineffective. As economic pressures increased, a large number of inshore fishermen "began to circumvent the controls" by misreporting the amount of groundfish caught, where the fish were caught, and which species was caught.

Control measures failed to protect the overfished stocks, leaving the inshore sector in crisis. The EA program designed to regulate the offshore sector had achieved a degree of success in that regard. However, it generated a need to maximize the economic return on the given allocation, increasing dumping, discarding, and highgrading of fish at sea. Estimates suggested that an amount equal to or greater than three times the recorded catch had been destroyed by turfing unwanted and dead fish back into the ocean.

The Harris and Hache Task Force reports both found that while detection of illegal activities was difficult for management and enforcement personnel, the penalties imposed once offenders were brought to justice were ineffective deterrents. Both reports recommended that observer coverage on the domestic offshore fleet be increased to 100 per cent and that penalties be more severe for such activities as overfishing, misreporting, dumping, discarding, and highgrading.

It may well be difficult to comprehend why fishermen engage in destructive fishing practices. Why do they not only cheat the system, but cheat themselves out of the resource that sustains their livelihood? Perhaps one answer, found in Dunne's report of 1990, is that over the past two decades, "many [fishermen] have lost a long-term commitment to the fishery and have turned to the short-term expediency of getting as much fish as possible regardless of the implications." It could be that independent fishermen and those employed on offshore trawlers do not entirely choose this course of action. They may be responding to mounting economic pressures that have increased tremendously over the past twenty years, fuelled by the need to continually upgrade fishfinding and catching capacity in an effort to survive, which means competing. For many of them, it may not be a matter of choice.

However, to lay the blame only at the feet of fishermen would be unfair. The Minister of Fisheries and Oceans and the DFO have been entrusted the stewardship of Canada's ocean resources. While most fishermen may publicly espouse the virtues of non-destructive fishing practices and express the desire to continue a tradition which showed more respect for marine resources, few have faith in the groundfish management system within which they are supposed to work because it has failed so miserably. They want fair and equitable management that will establish principles and guidelines to save the fishery from its self-destructive tendencies.

Chapter Six

False Data, False Ethics: Impact of Government Policies on the Fisheries

Despite good intentions, government policies designed to guide the Atlantic groundfishery to prosperity and stability have fallen short of the mark. Programs instituted by government have typically proven ill-advised and ineffective and have often intensified the cycle of boom and bust in the business. Industry leaders have had to contend with unsound policies and they are quick to remark that the fishery has survived not because of government policies, but in many cases despite them.

The ocean's resources required little attention until the middle of the twentieth century. The demands placed on the northwest Atlantic groundfishery by a stable market and a leisurely pace of expansion in technological power created the impression that the only threat to the health of fish stocks would be natural or environmental factors. Even with dramatic improvements in fish catching capability following World War II, there was little immediate concern from fishermen or government for regulation of the Atlantic groundfisheries. As catches increased and fishermen were able to pursue new and productive grounds, fish resources off Canada's Atlantic coast seemed beyond the realm of overexploitation.

Since 1944, the federal government has become more involved in the Atlantic groundfishery, with the aim of bolstering the industry to compete internationally and to take advantage of the market in the northeastern United States. A 1944 study conducted by S. Bates of Dalhousie University for Nova Scotia's Royal Commission on Provincial Development and Rehabilitation, proved to be a milestone in the growth and maturation of the groundfishery. The report, which was conducted to determine ways to revitalize the economy of rural Nova Scotia, pointed out that the domestic fishery was underdeveloped at best but had the potential to be a key player. It was recommended that capital, both public and private, be made available to offshore and inshore fishing interests to assist in the establishment of a competitive fresh and frozen fish industry in Nova Scotia. Included in the report was the call for government assistance for the construction of large draggers, and subsidies for processing plant modernization (to help with building retail and wholesale storage and new refrigeration capacity). For the inshore small boat sector, Bates called on government to provide loans for the construction of small draggers and longliners, similar to the assistance to be given larger operations.

It was apparent that new technologies and increased fishing capacity would put Canada on a firmer competitive footing and boost the lagging economy of Atlantic Canada, so government pulled out all the stops, offering loans and subsidies to competing segments of the industry. Even before the Bates report, the federal government had supported industry expansion. In 1942, corporations were offered assistance to construct trawlers over 22 m (72 ft.) in length and incentives to convert existing vessels to trawl technology. The package was improved in 1943, with subsidies of $165 per tonne offered to companies constructing large trawlers and allowance for accelerated vessel depreciation and write-off. Because of size and scale of operation, the offshore sector was seen by government to offer the best hope of re-establishing Canada's reputation as a seafood producing nation, lost during the war. Subsidies were also offered to encourage construction of inshore vessels designed to fish the lucrative grounds further from shore. In response to the Bates report, the Nova Scotia Fisheries Loan Board was established in 1947, with federal assistance providing loans to fishermen for the construction of both draggers and longliners.

The assistance for inshore vessel construction was most readily accepted by fishermen in southwestern Nova Scotia, leading to the development of a powerful inshore dragger fleet that could supply the fresh and frozen fish markets of the New England states. Conditions for federal subsidies called for the adoption of the newest technological advances to achieve the objective of increased catching power. Federal and provincial plans providing for capital assistance were modified over the next few years. In 1953, a significant change was made in the regulations which made federal subsidies conditional once large trawlers were affiliated with processing companies. The result was the continued development of the domestic offshore fleet throughout Atlantic Canada. The Newfoundland offshore fleet doubled in size from 1957 to 1966, while the already powerful Nova Scotia offshore fleet increased by one-fifth.

After World War II, when powerful distant water fleets were developed largely by European countries that fished around the globe, coastal states on both sides of the North Atlantic realized that unregulated fishing effort by these roving offshore trawler fleets posed a real danger to the health of groundfish stocks. Calls were made for government intervention and regulation. Nations fishing the northwest Atlantic in 1949 formed a regulatory body called the International Commission for the Northwest Atlantic Fisheries (ICNAF). Its mandate was to undertake research and bring order to the fishery by establishing minimum mesh sizes for trawl and by setting an annual total allowable catch with allocations to member states for the various stocks.

However, ICNAF proved terribly impotent. Although it set regulations on mesh size and other gear controls, it was unable to enforce these measures. During the quarter-century following World War II, the fisheries of the northwest Atlantic underwent a major transformation, and groundfish catches doubled. Only one-third of the total catch was taken by foreign fleets in 1950; that portion grew to nearly two-thirds by 1970. This increase was largely the result of the increased fishing effort and power of the fleets of the former U.S.S.R., Poland, and former East and West Germany, all of which were equipped with the most recent freezer-trawler technology. Despite the establishment of quotas for ICNAF members in 1971, the regulatory body was unable to enforce them, and the Canadian catch of Atlantic groundfish dropped to its modern low of 418,000 tonnes (matched in 1992, the year of the northern cod moratorium).

In 1979, ICNAF's successor, the Northwest Atlantic Fisheries Organization (NAFO), established closed areas and seasons to offer spawning and juvenile stocks a respite from the constant fishing effort by aggressive fishing fleets. The reputation of NAFO has suffered over the past decade because it has failed to respond to the aggressive overfishing efforts of Spain and Portugal that target turbot and other groundfish species just outside Canada's 200-mile limit.

During the 1950s and 1960s, foreign fishing went largely unmonitored and inshore catches were in decline, while the offshore sector's catch in Atlantic Canada increased steadily. In 1966, the offshore catch exceeded the total inshore take for the first time. The preferential treatment shown the offshore by way of loans and subsidies encouraged its unrestrained growth and overcapitalization through the postwar years.

Independent inshore fishermen showed a reluctance to invest in modern boats and gear, although Lamson and Hanson's collection of case studies shows that 125 longliners and 34 draggers were built in Nova Scotia between 1947 and 1960 with the assistance of government funding. According to Mitchell and Frick, by 1968-69 a total of 285 inshore vessels, between 25 and 100 gross registered tonnage, were either constructed or acquired in Nova Scotia with government assistance. Most of these were small boat draggers. The same report records a change in the guidelines for support of construction of inshore vessels that was instituted in 1965. Assistance for boats 11 m (35 ft.) in length—down from the previous minimum of 14 m (45 ft.)—was allowed, along with a minimum gross registered tonnage of less than 25 tonnes. In the three years immediately following this change, a total of 192 boats were purchased in Nova Scotia. At the same time as subsidies were becoming more accessible, uncertainty about the availability of fish stocks for the inshore fleet continued, until the extension of Canadian jurisdiction in 1977. The inshore accepted the extension as a signal of future stability, and the inshore dragger fleet started to expand with a vengeance.

While federal and provincial governments provided financial incentives, the Industrial Development Branch of the federal Department of Environment was instrumental in supplying essential technical support for the enhancement and expansion of the domestic fleet. Reviewing the work of the Industrial Development Branch from 1965 to 1975, a federal publication titled "Ten Years of Fisheries Development, 1965-1975" remarks on the progress: "Technological advances have been swift and their impact tremendous. The development and application of finding, catching, and handling fish have modernized and expanded the fishing fleets, a process spurred by the ever-increasing competition from abroad in exploiting the available stocks. At the same time, the shore-based [fish processing] sector of the industry has steadily improved its performance and increased its output of quality products."

A total of 830 projects were carried out over that ten-year period from 1965 to 1975 with the greatest concentration placed on the development of newly tested technology and technological improvements imported from other parts of the world to assist the Atlantic Canadian fishing industry. Though emphasis was placed on improvements in fishing methods and techniques, assistance was also provided to the shore-based sector to improve processing efficiency and to help with development of new fish products. The Atlantic offshore sector received a considerable portion of the attention, but the inshore dragger fleet received a substantial boost by the work of the Industrial Development Branch.

The goal of the Industrial Development Branch was to improve Canada's competitive position among fishing nations. It directed considerable effort toward adoption and modification of a trawl net design suitable for use on the Atlantic coast. The Atlantic Western Trawl, a four-panel trawl, had been developed on the Pacific coast by federal researchers and technicians and was considered to have potential as a replacement for the two-piece trawls in use on the Atlantic coast during the early 1960s. The Western Trawl proved capable of catching more haddock, pollock, and cod than other nets in use and was less susceptible to damage from towing over the ocean bottom. The same research team which had been involved in the development of the four-panel trawl on the west coast supervised the necessary modifications to make it a powerful tool in the hands of inshore dragger fishermen and offshore trawlermen.

The Industrial Development Branch was also instrumental in development of the stern trawling method used by the inshore and offshore fleets. This method greatly improved the efficiency of operation over side trawling and gave the crew greater safety in stormy weather. With the handling of the catch at the stern, space is used more efficiently as well. With these improvements in fish catching and fish handling capability, both the domestic inshore and offshore sectors were ready to compete with foreign fishing interests.

The overfishing of stocks outside Canada's 12-mile limit through the late 1960s and early 1970s convinced federal authorities of the need to gain management control of the continental shelf to bring fishing effort more in line with a sustainable level of harvest. A collapse of Atlantic groundfish stocks in 1974 forced the federal government to put the development of domestic capacity on hold and to explore ways to preserve stocks. A 1976 federal paper titled "Policy for Canada's Commercial Fisheries" outlined how harvesting effort could be controlled by a sophisticated licensing and quota regime designed to reach a new management goal of an optimum sustainable yield of groundfish, to replace an earlier target which allowed for a maximum sustainable yield, or a yield which, if pursued, could result in collapse of the groundfish population.

With the 1977 extension of Canadian jurisdiction to 200 miles from shore, the federal government gained control over resources on the continental shelf. The stage was set for a quota management system for groundfish stocks. The federal paper "Policy for Canada's Commercial Fisheries" determined that a quota management system should be a key to the restoration of overfished stocks and assist in providing a stable and sustainable resource base. In fact, a quota management system was required under the Law of the Sea (a United Nations authority and established convention used by independent states to determine principles of marine access and right) as a condition of the extension of Canadian jurisdiction. Under this authority, Canada was obliged to provide scientific support for the establishment of total allowable catches for individual groundfish stocks within its economic zone and then make available that portion of the resource not required by the domestic fleet to those nations with an established history of involvement in the fishery off Canada's coast, such as Russia, Cuba, Germany, and Japan.

The total allowable catches or quotas were set by a committee of regional director generals along with the Minister of Fisheries and Oceans who used scientific information, provided until recently by the Canadian Atlantic Fisheries Advisory Committee (CAFSAC). This background information was then made available to the various groundfish advisory committees that recommended changes to the Atlantic Groundfish Advisory Committee (AGAC). This umbrella committee, representing the federal government as well as industry groups, each year decided on the split of TAC between inshore and offshore, mobile and fixed gear, as well as the various subdivisions within these sectors. In 1992, CAFSAC and AGAC were replaced by the newly created Fisheries Resource Conservation Council (FRCC), established to broaden the advisory mandate from counting fish and setting the TAC to a mandate which also considered natural and manmade factors that threatened conservation of the stocks.

The supplanting of maximum sustainable yield (MSY) with optimum sustainable yield (OSY) in 1977 meant that a more conservative level of harvesting effort would be pursued, which would, in theory at least, provide for adequate replenishing of groundfish stocks while taking into account mortality caused by environmental factors and predation. While the previous MSY target permitted harvest rates approaching 35 per cent, the OSY level, based on calculation of a rate of fishing mortality termed $F_{0.1}$, would allow for approximately 20 per cent of the biomass of any gorundifhs species, or two out of every ten available fish, to be taken annually. The $F_{0.1}$ formula was established in rocognition of the fact that for each stock there is a level of industry fishing effort beyond which catch rates begin to decline markedly for all participants, due to increased competition. For cod, this target level of annual harvest, as produced by calculation of the $F_{0.1}$ formula, is more precisely pegged at 16 per cent, while for American plaice the level is 20 per cent. Fishing effort above the OSY level on a prolonged basis results in lowere returns to the industry because a protion of th sotck is taken before it reaches its full growth potential. Computation of an acceptable total allowable catch using the $F_{0.1}$ formula requires data inputs on the biomass, mortality, and recruitment of new memebers to the stock.

The biomass of a stock refers to the number of fish in the ocean multiplied by their individual weights. Since scientists have no way of accurately counting every fish born to a stock in a particular year, they draw upon fish landings or catches as one of several key elements in their calculation of stock biomass. They examine commercial catches to determine the age of the fish; they weigh and count the fish to determine the weight and number of fish in that age class. From this information they can roughly determine the biomass of a particular stock born in a given year. For example, if the fishery were to catch 50,000 three-year-old cod from one particular stock in 1993, and another 50,000 four-year-old cod from that same stock in 1994, then there had to have been at least 100,000 cod in the stock born in 1990. A more accurate figure would be available by adding to that calculation the catch of five-year-old cod caught in 1995 and so on. The biomass is then calculated by multiplying the total number of cod from that particular stock born in 1990 by their individual weights.

Additional data useful in the calculation of biomass is gathered by research cruise ships operated by DFO researchers. These cruises involve the yearly towing of a bottom trawl through a stock's range, whether it is inshore or offshore ground. The same area is trawled at the same time each year to provide consistent spatial and temporal coverage. While researchers acknowledge that fish habits and patterns of migration may vary from year to year, these sometimes coarse results are nevertheless considered valuable when added to the information on biomass gathered by other means.

In order to calculate the mortality rate of groundfish stocks to input into the computation of total allowable catches, scientists have to take into account not only the recorded fish landings but also the discards of each species. Discarding generally occurs because fishermen do not have a market for small fish or are prohibited from landing them because of size restrictions. The rate of discards is difficult to assess accurately because it is carried out in secret and goes largely undetected. A factor for natural mortality arising from environmental conditions and predation must also be entered. The total of mortalities gives scientists an idea of the amount of fish leaving the stock.

Recruitment, another consideration in the calculation of TACs, is a measure of the number of young fish joining the stock which have attained sufficient size to be legally harvested. The total number of fish reaching the legal size for harvest multiplied by the average weight gained by each fish indicates the total weight to be added to the biomass of the stock.

The federal scientists' confidence in the $F_{0.1}$ formula was bolstered following 1977, when stocks benefited from relatively good recruitment. With the foreign fleets gone, the number of cod and other groundfish off the Atlantic coast increased, showing signs of stable growth from 1978 through to 1981. Encouraged by federal subsidies and optimistic predictions from politicians and scientists alike, the domestic fleet, particularly the offshore, underwent a major expansion in fish catching and processing capacity. Federal scientists were predicting that the northern cod catch, which had been reduced to 139,000 tonnes by 1978, could be increased to 350,000 tonnes by 1985. Similar rosy forecasts were being made for most Atlantic groundfish stocks with a total catch of 1.1 million tonnes of groundfish being predicted for the entire region for 1987.

When business downturns in the early 1980s threatened the inherent instability caused by government hand-outs, the offshore sector was left insolvent. Spiralling interest rates and a softening of fish markets undermined expansion and the offshore sector crashed. Led by Ottawa, a government bail-out of the offshore sector was arranged. As part of the deal, the large companies were granted company quotas called Enterprise Allocations (EAs) in 1982. Under the EA program, offshore companies were given a portion of the overall total allowable catch of groundfish and allowed the freedom to harvest that fish any time before the year's end. Fisheries managers reasoned that the assignment of company quotas would allow offshore companies to reduce overhead, helping to make their operations more profitable. The offshore sector was granted nearly half of the total allocations of Atlantic groundfish. The inshore sector, on the other hand, remained on a competitive quota; each boat was left to rush for its portion of the remaining quota not assigned to the offshore. The new arrangement did not sit well with the inshore, which claimed that the large companies had been given economic power over the region's natural resources with the uneven distribution of quota.

The apparent success of the quota management system during its first few years (1978-1981) could be attributed to the hard line maintained by fisheries managers. Conservative TACs were established to allow stocks to recover from overfishing, and efforts were made to restrain frenzied expansion of fishing capacity. In 1979, a freeze was placed on entry to all sectors of the fishery, and in 1981, the fixed gear was placed under quota management for all groundfish stocks. In 1982, a sector management system was introduced for all vessels under 20 m (65 ft.) in length, restricting boats to registration in one of three zones—the Gulf of St. Lawrence, Scotia-Fundy, or Newfoundland. Sector management was established to restrict the movement of vessels and thus reduce the likelihood that congregating or spawning stocks would come under intense pressure from mobile fishing efforts.

Despite the good intentions of the quota management system, it was seriously flawed. The system did succeed in dividing the quota among gear types and boat sizes but also caused division among fishermen over the fairness of their groundfish allocations. Conflict between industry sectors and interference from politicians working to protect the interests of the fishermen and sectors they represented derailed the quota management system. Despite evidence that the resource was no longer responding to management measures and that high catch rates were suppressing stock growth, a series of fisheries ministers, under duress from fishermen's groups and lobbying from powerful offshore companies often set quotas at levels above those recommended by scientists. This seemed to occur more frequently as shortages of fish became more pronounced. Fish stock health suffered as a consequence. Fishing efforts of both the inshore and offshore were not adequately constrained by the liberal quotas. In 1989, a shortage of quota forced the Scotia-Fundy inshore dragger fleet to endure its first serious shutdown, with closure of the season coming at the end of June. Processing plants located primarily in southwestern Nova Scotia and dependent upon fish caught by this dragger fleet were slowed by the early closure; the plant workers were laid off during the downtime.

The early closure brought home the realization that the inshore dragger fleet could actually chase the stock down to the last fish. It became apparent that the stocks would succumb before the fleet was forced by economic constraint to tie up. The solution by government was an inshore boat quota similar to the offshore Enterprise Allocation system. This was the primary recommendation of the 1989 Hache Task Force, established to make changes to the Scotia-Fundy management scheme. After a year of cursory consultations between representatives of the inshore sector and government, the boat quota system was imposed on the under 20 m (65 ft.) dragger fleet in 1991. Boat quotas had already been introduced for vessels between 20 m and 30m (65 ft. and 100 ft.) in length in 1988.

The unanticipated rapid increase in harvesting capacity of the inshore and offshore sectors of the domestic fleet after 1977 challenged the successful operation of the quota management system. The growth in offshore catching capacity was deflected by the establishment of the Enterprise Allocation system, which removed the advantage of adding harvesting power. The inshore sector, comprised of hundreds of independent interests involved in a competitive fishery, proved much more difficult to control.

First in 1978, and again in 1980 and 1987, the federal government announced a moratorium on the issuance of more licences to pursue different species of groundfish. Licences that had been inactive would be retired if the owner was unable to demonstrate its vital importance to his operation. This generated the attitude of "use it or lose it" among fishermen. Suddenly, these inactive licences carried renewed value. To justify their retention, a considerable number of fishermen built boats to carry the licences. Later, some of these licences and boats were sold to others interested in entering the fishery. The restriction on the issuance of licences caused a tremendous expansion in the Scotia-Fundy inshore fleet, in particular in the dragger sector, where four hundred powerful new vessels had been added over the decade. Rather than buying out the licences, the federal government permitted fishermen to invest in larger boats. Some, who had not seriously considered entering the fishery prior to the moratorium, did so. The management measure of licence-issuing moratorium, designed to hold or reduce capacity, had backfired and further destabilized the relationship between resource availability and harvesting capacity.

An increase in fleet carrying capacity resulted from another management measure introduced in 1981, which allowed vessel owners looking to replace their older craft an incremental increase of about 1 m (5 ft.) in length and a 10 per cent increase in fish hold capacity. The new regulation was designed to slow the rate of increase of vessel size without limiting it to zero. This management measure also backfired as managers had not predicted the extent of improvisation by fishermen. Fibreglass, a stronger and more flexible material than wood, was used to form the hull of the new vessels because it allowed for the construction of wider beamed vessels with much greater carrying capacity. These squattier vessels, or "jumbos" as they came to be called, met the legal requirements set out by the federal regulators and were soon all the rage. The design of the jumbos was especially attractive to dragger fishermen committed to a higher volume fishery.

In 1989, the federal government realized their blunder and imposed a new licensing and vessel replacement policy which froze carrying capacity. The adoption of a three-dimensional or cubic capacity measurement for vessels prohibited further increases and effectively plugged the loophole that had allowed the construction of the wider beamed jumbos. This move came far too late as the overwhelming majority of draggers had already been converted to the jumbo design.

While federal biologists were initially confident that the domestic fleet quotas and the regulations to restrict catching capacity would allow a rebuilding of stocks, there was a growing uneasiness in the ranks of the scientific community and among fishermen. The average size of fish was decreasing and most inshore fishermen were finding that catch rates were plummeting. The inshore fishermen of Newfoundland and Labrador and the Gulf of St. Lawrence were, for the most part, the first to call attention to falling catches. These fishermen had fished the same waters for generations and were well situated to measure the downturn. Like fishermen along the rest of the Atlantic coast, they had been subjected to poor catches because of the intense overfishing in the 1960s and 1970s and had hoped for better catches following 1977. But the stocks did not rebound as expected. The optimistic forecasts made by scientists upon extension of Canadian jurisdiction had been well received in 1977, but by the mid-1980s they were sounding rather hollow to those in the inshore. None of the target TACs granted to the Newfoundland and Labrador inshore were ever reached during the 1980s—there were simply not enough fish. Yet federal fisheries managers blamed the poor inshore catches on deteriorating environmental conditions and continued to overestimate the size of the stocks.

Apprehension was also growing within the scientific community. Though there was relatively good recruitment of groundfish stocks between 1978 and 1981, and though the northern cod stock had tripled in weight between 1976 and 1988, there had been a dramatic downturn. The *Sou'Wester* reported on May 15, 1989, that scientists found the 1988 northern cod stock was only two-thirds the size advised by the Canadian Atlantic Fisheries Advisory Committee in 1987. CAFSAC attributed the small stock numbers to higher fishing mortality than had been expected; in 1987, the fishing mortality was perhaps double the $F_{0.1}$ target of about 20 per cent of the harvestable stock. Although Fisheries Minister Tom Siddon reduced the 1989 TAC for northern cod from 266,000 tonnes to 235,000 tonnes, the TAC reduction fell short of the 125,000 tonnes that CAFSAC scientists had recommended. These scientists had noted some ominous signs from the data collected by research ship cruises conducted in the fall of 1988. The politicians and managers concerned about the impact of quota reductions on the fishing industry failed to set tough quota rollbacks. The stocks were again exposed to excessive harvest rates.

In their 1987 report, CAFSAC biologists cited four of the signs that indicated trouble for the health of northern cod stocks. First, commercial catch rates had not increased since 1985, despite increased effort exerted by the offshore sector. Second, research vessel survey tows in 1987 and 1988 conducted by DFO scientists, indicated that the stock had fallen back to levels typical of the first few years of the decade. Third, a critical examination of the management since 1977 indicated that fishing mortality rates had consistently been too high—and with catch rates about twice those recommended—pressure from the domestic fleet had surpassed the limits of sustainability of the resource. Finally, the 1988 examination of the age profile of the stock showed that there were too few of the older, more reproductively potent fish present in the commercial and research catches. Numbers of fish older than eight years declined sharply in both catches, indicating that they had been fished harder than previously estimated.

The CAFSAC team pointed out the 1987 and 1988 commercial and research vessel data had clearly shown that earlier assessments of stock size had been much too generous while the growth and abundance of northern cod had been overestimated. Although different methods of calibration were proposed by inshore fishermen, the CAFSAC scientists were careful to point out that it was the new data inputs, and not any change in the model for computing stock size, that had brought about the re-evaluation of size of the northern cod stock. They suggested that the causes for reduction in size might be a lower number of spawning stock and unauthorized increases in the rate of commercial harvest.

The triad of scientific, administrative, and enforcement functions carried out under the Department of Fisheries and Oceans is fundamentally dependent upon accurate estimates of stock health. The data inputs collected by scientists are used to set sustainable levels of fishing effort. The advice offered by biologists is instrumental in the establishment of regulations designed to protect stock health. Over the years, the Canadian government has made tremendous annual investments in scientific research, administrative policies, and enforcement measures necessary to make the quota management system work, demanding inordinate commitment from government and non-government participants to shape a system that could secure industry and resource viability. The energy, time, and financial resources dedicated to the task of Canadian fisheries management is unparalleled. Yet, the management system failed one of its primary tenets—resource sustainability. For example, since 1983 total allowable catches of groundfish established by government and industry have fallen steadily. Groundfish stocks from Labrador to the American border are in poor or failing health. Falling catches experienced by the Newfoundland and Labrador inshore and increased effort required by the offshore trawler fleet to catch northern cod as early as 1984 should have indicated trouble in that fishery. The inshore mobile fleet in the Scotia-Fundy region, fishing from a sector quota assigned to the dragger fleet, was catching its entire yearly quota before the 1989 season was half over. The problems were Atlantic-wide because the quota management system was malfunctioning.

One of the primary reasons for the failure of the management system was the overriding preoccupation of the DFO with establishing, allocating, and enforcing quotas for the various sectors and fleets involved in the Atlantic fishery. The administrative priority of "management by quotas" has strained manpower and fiscal resources, leaving very little attention for other aspects, such as holistic study of the ecosystems on which groundfish are dependent and the impact of various harvesting methods on the target species and its habitat. When the federal budget of 1985 slashed DFO expenditures, it responded by closing the Marine Ecology Laboratory in Halifax a year later. Further, while an entire division within DFO, called Resource Management, was established to allocate resources and determine quotas, today, there is not a purely management division charged with ecosystem research and protection as it applies to harvesting methods and technology. The effects of the misdirected effort toward stock allocations are still being felt as scientists struggle to fill the gaps in our basic knowledge of the marine environment by conducting long overdue pure research on the ocean's systems.

The time and resources dedicated to management by quotas offered little opportunity for a dissenting view and blinded managers to the fact that the system left aggregations of spawners and juveniles vulnerable to attack by the aggressive mobile gear sector. By 1989, the problems of the Atlantic groundfishery were headline news. Stocks were crashing and the industry was in turmoil. How could the fishery be in crisis again when only ten years earlier the industry and the Canadian government had been given carte blanche to undo the damage created by years of foreign and domestic overfishing? How could it be that the Canadian system of fisheries management, praised internationally as the head of its class had failed to foresee the collapse of the resource? How could a heavily regulated fishery be allowed to drive groundfish stocks to the brink of disaster?

With the economy of the Atlantic region so destabilized, it was understandable that the public would want to lay blame. In a 1989 survey by Corporate Research Associates Inc. of Halifax, almost half of the 1,500 people surveyed indicated they thought the federal government was responsible for the trouble in the fishery. The Trudeau Liberals, in power for all but one year (1979) of the sixteen-year span from 1968 to 1984, had set the stage for disaster. It was under Prime Minister Pierre Trudeau that the offshore was rescued from bankruptcy with a bail-out in the early 1980s, and it was during this era that federal grants were handed out willy-nilly to almost any interest expressing a desire to build up Canada's fish processing and harvesting capacity. Little thought was given to the eventual impact on the resource of this increase in fishing and processing power. The federal Tories, for their part, did little to arrest or even slow the free fall. Prime Minister Brian Mulroney's Government (1984-1993) merely presided over the burial of the

Atlantic groundfish fishery. Neither governing party had a corner on the mismanagement of the fishery.

The federal government did, however, play a reluctant role in bringing the groundfish crisis to a head. In February 1989, Minister Tom Siddon—under increased pressure from all segments of the industry and with a confession from his own scientists that they had grossly overestimated the health of northern cod stocks—commissioned an independent review of the state of the northern cod stocks and five months later, a review of the state of the Scotia-Fundy groundfish fishery. The independent review, conducted by the Northern Cod Review Panel, of the DFO's research methods as the basis for their advice was ordered by Siddon in February 1989. Leslie Harris, President and Vice-Chancellor of Memorial University of Newfoundland, chaired the Northern Cod Review Panel and reported in February 1990. The Harris Report, as the review was called, confirmed that earlier scientific advice had been overly optimistic and that it had been based on projections which indicated the 200-mile limit would allow cod stocks and prosperity to return to the Atlantic coast. This fever of excitement and optimism had, of course, infected all those associated with the cod fishery. The Harris Report also suggested that the mathematical modelling had been inadequate and that it had been fed by insufficiently examined data.

The review panel pointed out that the $F_{0.1}$ formula was flawed because it did not recognize the need to maintain a sufficient number of older aged spawning members in the cod population. The $F_{0.1}$ formula simply measured the biomass comprised of all year classes of exploitable age and paid no particular attention to the health of the sexually mature age group, which would consist of individuals six years and older. It is the older female fish (10-15 years and older) which are most productive. In hindsight, this is a factor which should have been considered in calculations used to determine the level of harvestable biomass and reflected in the total allowable catches of cod.

While the $F_{0.1}$ model was flawed, the data used for the calculations was also found wanting. The panel stated its belief that both the research vessel data and the commercial catch data were incorrect. The research tows are conducted generally in July when the codfish are in a migratory mode and scattered throughout a large quadrant of the ocean, including inshore and offshore grounds; therefore, the results of the research trawls are guaranteed to carry significant variables. The survey tows cover only a very small portion of the entire range of cod stocks under study, and so they operate on a trial-and-error basis. Also, cod seek out ideal conditions of water temperature, salinity, and availability of food. But scientists intentionally disregard these ecological factors to avoid bias in their findings. They then expose their research to considerable error, which may vary from less than 10 per cent to 50 per cent or more. They run the risk of returning with misleading information that would only be righted by additional survey of the stock.

Consistency in the commercial catch data was even more problematic according to the Harris Report. Scientists using the $F_{0.1}$ formula had wrongly accepted the premise that improvements in catch technology over time did not effect the catch rate of offshore trawlers used to collect commercial catch data. Catch per unit effort (CPUE) refers to the weight of fish removed by a definable fishing gear over a unit of time and is a key measurement used by scientists. With no improvements in catch capacity or power, an increase in CPUE would imply that stock health is improved because fish are easier to catch and, therefore, stocks must be in good shape and able to sustain a greater fishing effort. Scientists overlooked the tremendous improvements in catch technology and catch power. When CPUE numbers remained relatively stable in the face of these improvements this should have signalled that fish were more difficult to catch and stock health was in decline.

Interviewed by the Standing Committee on Forestry and Fisheries of the House of Commons, Harris explained how the index or formula used to estimate cod stock abundance from commercial catch data was warped. "...We maintain that the technological advance has been so great that the index is skewed out of recognition. In other words, a unit of effort—an hour's fishing—in 1990 is worth perhaps three hours of fishing in 1975 or 1980, simply because the technology has improved so greatly. The quality of the fishing gear, the electronic capacity to seek and find, the speed and power of vessels, their capacity to fish in ice conditions, the skill and experience of skippers and crew—all of these things have changed enormously, but none have been taken into account or reflected in the construction of the index. So we believe the index of abundance derived from that source is not a very realistic one."

The Harris Report goes on to point out that the index of abundance is not realistic for another reason. With the fleet able to detect swimming groundfish with the array of electronic devices available, fishermen save time by not setting their gear where the prospects for a bad catch are obvious. That effort is, instead, saved for waters attracting aggregations of commercial stock with the result that the catch per unit effort is maintained, despite overall reduction in groundfish numbers. In addition, the Harris Report confirmed suspicions held by Canadian fishermen that foreign vessels were underreporting catches made inside Canada's territorial waters. Reliable sources reported to the panel that even on vessels carrying Canadian observers, underreporting of up to 25 per cent of the catch was "not all that uncommon."

Incorrect suppositions about the abundance of northern cod stocks were applied to assessment of the health of other Atlantic groundfish stocks. Biomass calculations for determining TACs for Gulf of St. Lawrence and Scotia-Fundy stocks were also based on input from blind surveys conducted by research vessels and on statistics on commercial catches gathered by DFO field workers. On several occasions in 1988 (March 8, May 17, October 14, and December 20), *The Guardian Newspaper* reported that evidence produced in courtroom proceedings has shown the raw data provided by fishermen is purposely inaccurate because captains respond to management measures, such as catch quotas by misreporting or underreporting their landings. Neither random surveys or commercial catch data can provide a consistent and workable estimate of stock abundance, greatly weakening the reliability and usefulness of this key index of abundance in fishery management. The findings of the Harris Report ultimately cast a shadow on the research methods and data used by DFO scientists.

Not long before the release of the Harris Report, Siddon responded to questions in the House of Commons on October 2, 1989: "...in most respects fishermen in Atlantic Canada are enjoying a year which is prosperous and adequate for their needs in many parts of Atlantic Canada, but not as abundant as it had been in the last three or four years in relative comparison." He suggested that the only crisis the fishery was facing was one of expectation. That same theme was repeated one week later by John Crosbie, Deputy-Minister of Fisheries and Oceans. The DFO failed to live up to its mandate as steward of the ocean and its resources by denying there was a crisis. The official word out of Ottawa, Halifax, Moncton, and St. John's was denial—"the fishery is not in crisis."

Siddon's comments provoked a furore from the Opposition in Ottawa and the industry in Atlantic Canada. Newfoundland Liberal MP Roger Simmons argued in the House of Commons on October 2, 1989, that "almost any fisherman, almost any plant worker in Atlantic Canada, can vouch for the fact that the Minister's statement is simply not true.... They are certainly not going to convince anybody involved in the Newfoundland fishery, either the inshore or the offshore deep-sea fishery, that there is no crisis. They are not going to convince any of those people that the fishery in Newfoundland is not on the verge of collapse." In *The Chronicle-Herald* of October 5, 1989, Glenn Wadman, Vice-President of the Independent Seafood Producers of Nova Scotia, called the Minister's same statement "staggering and unbelievable," pointing out that fish plant operators in Nova Scotia were concerned for the future of their businesses and their personal possessions.

The extent of denial by Siddon of a groundfish collapse was astounding with the January 2, 1990, announcement of northern cod quotas under the 1990 Groundfish Management Plan. The advice of his own biologists, who admitted they had overestimated the northern cod stock by a full 100 per cent, was to cut the recommended northern cod quota from 235,000 tonnes to 125,000 tonnes. Instead, he reduced the number to 197,000 tonnes. Siddon did not wait for the recommendations that came one month later in the Harris Report. This drew harsh criticism from Wilfred Bartlett, Vice-President of the Newfoundland Inshore Fishermen's Association, who was quoted in the *Sou'Wester* of January 15, 1990: "It's the same old political game that is being played—when people are concerned or kick up a little stink, appoint a commission to look into the problem and when the report is finished, ignore its findings."

On January 22, 1990, Siddon was supported by Prime Minister Brian Mulroney who reiterated his commitment to aid for communities hard-hit by the failing Atlantic groundfishery but denied there was a crisis. Before taking his seat in the House of Commons, he was persuaded by Opposition members to acknowledge a crisis in the Atlantic groundfishery, though he indicated he would rather not focus on it. "...We have acknowledged that the Atlantic fishery is in a state of crisis, and we are cooperating with provinces, communities, and other national governments to make sure that we deal with that crisis. But we do not want the international investment community to believe that because there is a serious problem in the fishery that the economy of Atlantic Canada is to be scorned because it is a basket case. It is not a basket case. The economy of Atlantic Canada shows areas of important strengths. There is an economic future in Atlantic Canada. We do not want doom and gloom. We want to bring about greater prosperity."

Siddon's successor, Bernard Valcourt of New Brunswick was appointed by Prime Minister Mulroney in February 1990. He quickly disappointed the Newfoundland inshore sector by rejecting the prime recommendation of the Harris Report—that northern cod quotas be reduced immediately. Valcourt said he shared the view that the stocks were in decline but denied the appropriateness of a reduction in the 1990 quota, considering the hardships it would inflict upon the industry. He also failed to prescribe the recommended cuts in the 1991 Groundfish Management Plan, cutting the total northern cod quota to 188,000 tonnes but denying the inshore sector's request that a further 20,000 tonne cut be made with the reduction to be shared by the offshore. Valcourt said he wanted to strike a balance between saving fish and saving jobs. This proved to be a very myopic approach.

Responsibility for the Fisheries and Oceans portfolio eventually fell to Newfoundland MP John Crosbie, who was sworn in as Minister in April 1991. Fifteen months later, on July 2, 1992, he found himself announcing a two-year moratorium on the northern cod fishery and unveiling a recovery plan for a fishery obliterated by the closure. The announcement was made in the same hotel, where in 1990 he and Fisheries Minister Tom Siddon had disregarded calls for a shutdown of the northern cod fishery. Like those before him, John Crosbie moved too cautiously. He and the federal Conservatives paid the price for not slashing northern cod quotas in the fall 1993 federal election, when they were swept from power.

The official line from the DFO remained optimistic even during the worst years, when the offshore companies faced financial ruin; when the domestic fishery had to contend with failing groundfish stocks; and when the closure of the groundfish fishery appeared inevitable. The DFO chose to emphasize the overall strength of the Atlantic fishery—buoyed by the vitality of the shellfish sector—rather than openly admit that groundfish stocks were disappearing. Although peddling good times in the fishery served a purpose in not being alarmist, the perception portrayed by DFO did a disservice to the groundfish industry and those who depended on it. The ambivalence of the DFO frustrated the well-intentioned efforts of some sectors of the fishery, who reasoned with foresight that there was still time to turn the worsening situation around. Those concerned for the fishery's future tried in vain to persuade the department to rethink its short-term approach to the fishery, which placed a priority on maintaining a steady course and preserving jobs, rather than moving its direction toward conservation. Finally, the truth could not be denied. The crisis was real, and the opportunity for a smooth turnaround had passed.

Much of the mismanagement of government can be traced to its lack of understanding of the fish business and the resource on which it is based. Early on, government was headed for difficulty because it pursued contradictory agendas: one was designed to increase productivity and profitability of the domestic industry and the more recent tenet aimed to protect the fish stocks. Fishery management, which still relies heavily on information from government-produced stock surveys and scientific reports, often unwittingly based decisions on false data. Further, government did not recognize the significant role of modern harvesting technologies in the destruction of fish stocks and fish habitat. Perhaps most reprehensible, government ignored recommendations for quotas to be drastically reduced through the mid-1980s—a measure which may have spared the industry from the magnitude of its present crisis. Still, we cannot look to government as a scapegoat for a crisis in which so many played a part.

Chapter Seven

Politics of Alliance: Collusion and Coercion Among Government, Big Business, and Scientists

Few sectors of Canada's economy are as profoundly influenced by government policy as the fisheries. The fisheries are regulated from the landing of the catch aboard vessels to the finished product on the consumer's dinner table. Not surprisingly, industry stakeholders share a love-hate relationship with government. Every policy change by government is subjected to debate, analysis, and criticism from various sectors of the business. Industry is often the first to call for government intervention when the fishery faces a crisis but is quick to criticize measures as intrusions into its affairs. At the same time, the government assistance has been crucial to the success and even the existence of the offshore over the past three decades. Government attempted first to shield the industry from the threat of foreign overfishing and then to rescue the offshore from bankruptcy. Government also has been able to claim credit for the employment created and the boost this sector has given to the economy of rural Atlantic Canada. Throughout their history, government and big business have found mutual benefit from their developing alliance. Government has gained from being able to trumpet an offshore ready to compete on the international stage; the large companies have profited from the competitive advantage and security given them by government.

Federal fisheries policies, like other government policies, are largely determined by a number of high-level bureaucrats who report to the Minister of Fisheries and Oceans. These high-level DFO bureaucrats interact with representatives from the governments of the four Atlantic provinces and special interest groups from industry who, together, form a policy community which is most influential in determining fisheries policy, according to Whittington and Williams. Paul Pross suggests this inner circle, led by DFO, is generally insulated from input by individual fishermen and the general public, except in crisis situations, when past policy making is exposed to attack. While individual fishermen and even representatives of fishermen's organizations cannot hope to gain entrance to this sub-government structure, large corporations like National Sea Products, Fishery Products International, and Clearwater Fine Foods, with their size, influence, and organizational structures are capable of "interfacing" with the federal government on their own. These big business players have a corporate structure dedicated to nurturing a relationship with government that includes departments dedicated to long-term planning, research and development, and government relations. The Fisheries Council of Canada represents large processors and is considered a dominant player in this fisheries sub-government, influencing policy at the federal and provincial levels.

While the DFO would rather give the appearance of inviting input on policy formation from all sectors of the domestic industry, independent fishermen and independent fish processors realize they have little impact on government decision making. This is not news to anyone who has participated in the formal consulting process or has made representation to any of the countless public forums staged by the DFO. Fisheries policy making is government-driven and influenced only by those with lobbying power. The bureaucratic inertia inherent in the system effectively stifles independent input. Influencing this elite policy community requires a full-time public relations effort, something that only larger offshore companies in the industry can afford, and something that National Sea, for example, has invested in.

National Sea Products was fast off the mark with the declaration that Canada would extend its jurisdiction to 200 miles. They quickly went to work lobbying for a special place in the nation's future prosperity. A domestic offshore presence was required and National Sea was the company to lead the way. They pulled the right strings at the right time, capturing the imagination of federal politicians eager to get on with regaining power and position lost to foreign fishing nations.

Those pulled strings proved to be purse strings as well. Ottawa bent over backwards to respond to National Sea's proposal, financing the expansion with grants, subsidies, and loan arrangements. The government response was predictable when the prospects for development of an exploitable resource and the attendant prosperity were presented. The enthusiasm and fervour of political backing rode the wave of corporate anticipation. The larger and more immediate the prospects for gain, the more committed the political support became by loosening the reins on resource harvesting. Quotas, Enterprise Allocations, and exploratory licences were handed to the offshore, granting a disproportionately large share of available resources to this sector, with little thought of conservation of stocks or the impact on the fixed gear inshore sector. The objective was to Canadianize the east coast fishery at any cost. The silent agenda was that the prosperity that would accrue to rural Atlantic Canada would be translated into victory at the ballot box.

Federal money continued to flow even after National Sea and the offshore companies of Newfoundland found themselves carrying too much debt in the early 1980s. The Kirby Task Force recommended government bail-out of the large offshore company; the ground gained could not be lost. A restructuring agreement was signed in 1983 between Newfoundland and Labrador and the federal government, providing a total of $167.6 million in financing for Fishery Products International. In 1984, the government provided $30.6 million for National Sea. Fishery Products International eventually privatized the company by buying back the government's shares in 1987. Ottawa still maintains its original 19 per cent share of ownership in Nat Sea, according to an article of *The Chronicle-Herald* from June 18, 1991.

With such a large investment, Ottawa had, in effect, committed itself to long-term support of the offshore companies. Independent fishermen and fish processors openly declared that the federal government had invested so heavily in companies like National Sea Products and Fishery Products International that they could not afford to pull the plug. It was perceived that the door to the Minister's office was always open to the offshore. This unabashed alliance between government and the corporate offshore was a constant source of irritation for the inshore sector, which was convinced that Ottawa should not be propping up big business interests, such as National Sea Products, only to have them compete unfairly with small boat fishermen and independent processors. While these same small boat fishermen and independent processors were not above receiving government assistance for boat construction or plant expansion, they did not consider their fragmented voice to be particularly persuasive and were never apologetic for their stand. Leaders of inshore communities were not averse to confronting the DFO with what was perceived to be Ottawa's biased influence on fisheries management in favour of the large offshore companies.

The sentiment of the fixed gear inshore members is reflected in the following quotation from a "Draft Brief Prepared for Municipal Units in Digby, Yarmouth, and Shelburne Counties" to be presented to the House of Commons Standing Committee on Fisheries: "In a democracy such as ours it is recognized that justice must not only be done, but must be seen to be done. In this way the people can have confidence that the system is fair and that their rights and interests will be protected by government. This principle should also be applied to fisheries management. Fishermen should have confidence in the federal department whose mission is to protect and maximize their livelihood. It is the honestly held belief of fishermen that an incestuous relationship exists between DFO and the large fish processing companies. If this is a myth then it should be addressed. If, on the other hand, there is some truth to this, then it should be rectified."

The perception of favouritism did not deter Ottawa from granting a factory freezer trawler licence to National Sea Products in 1985. Federal politicians were willing to endure the ridicule of the industry, especially Newfoundland which was vehemently opposed to the licence, believing the FFT technology was destructive to groundfish stocks and their environment. The controversial announcement was made in Ottawa by acting Fisheries Minister, Erik Nielsen, and in Newfoundland by MP John Crosbie, then Minister of Justice and Attorney General. Inshore fishermen saw no justice, despite the promise in a DFO press release of December 8, 1985, that "the inshore small boat fishery and wetfish trawler fishery will continue to be the backbone of the Atlantic groundfish industry."

South Shore Nova Scotia MP, Lloyd Crouse, whose riding is home base for National Sea's operations, rose in the House of Commons on November 8, 1985, in support of Canadian ownership and use of factory freezer trawlers. "I believe that the time has arrived when we must reassess our priorities as they relate to quotas for foreign and Canadian fishermen within our 200-mile zone. FFTs will enable Canadians to find new job opportunities and open up markets now lost to foreign competition.... It is time we got on with the job."

The reciprocity between the corporate offshore and the government is best shown in the question of free trade with the United States. The 1988 federal election campaign had a slow start but heated up when the Free Trade Agreement (FTA) with the Americans—the cornerstone of Prime Minister Mulroney's economic plan—took centre stage. Canadian big business, such as National Sea and Clearwater Fine Foods, favoured the FTA because it promised to eliminate tariffs on a number of fish products, including ready-to-serve dinners, which would expand market opportunities south of the border. The general public, on the other hand, grew increasingly apprehensive that the free trade deal was designed primarily to serve corporate interests while sacrificing the protection of Canada's rural economies.

In recognition of the growing opposition, National Sea and Clearwater Fine Foods decided to enter the fray. They did their part by running full-page advertisements in support of free trade in regional and national newspapers, and top executives freely gave interviews in support of the FTA. The momentum of the federal campaign shifted as public opinion sided with the Conservatives and their promise of economic development and opportunity. The advertising campaign sponsored by corporations was considered a key factor in the turnaround, and the Conservatives won the election.

While the advertisements were sufficient evidence that the marriage between the corporate offshore and the federal government had not lost its sparkle, confirmation of the ties was given when an Elections Canada report the following spring established that Clearwater Fine Foods had contributed to the election campaign of Fisheries Minister Tom Siddon. The $2,000 contribution was made to his personal campaign in the Richmond, British Columbia, riding. Siddon told the press that he saw nothing improper about the donation. He said he rejected, "absolutely out of hand," any suggestion the contribution placed him in a conflict of interest. For those who were wary of the questionable conduct between the corporate offshore and the government, this was confirmation of their suspicions.

Clearwater President John Risley defended his company's support of the Conservative campaign and the FTA as simply good business. According to an article in the January/February 1989 issue of *Atlantic Business,* for Risley the election result and the subsequent signing of the agreement with the United States was a sign that the long-term security of the fishing industry was assured.

It was less than six months after the fall election of 1988 when offshore corporations, led by Clearwater Fine Foods, began closing their processing facilities in a number of towns and villages around Atlantic Canada, effectively walking away from communities and workers who had worked hard to make the company profitable. Clearwater closed the S. W. MacLeod Fisheries plant in Port Mouton, Nova Scotia, laying off 40 full-time and 150 seasonal workers. In May 1989, National Sea terminated the Lockeport operation, and temporarily closed three others in Nova Scotia and three in Newfoundland, affecting a total of 3,445 workers. In Lockeport, National Sea walked away from a facility built largely with taxpayers' money and was allowed to later sell or flip it to Clearwater at a fire-sale price. Clearwater, in turn, was able to dip into the public purse, grabbing another $3 million to finance the purchase, modification, and reopening of the plant. National Sea was also smiling after the Canso bail-out. The company was able to wipe out a $17.5 million debt on the plant and pick up a 2,000 tonne quota of whole shrimp at the same time.

Despite plant closures, the offshore corporations are busy. National Sea Products and Clearwater have benefited from the granting of exploratory and permanent licences to harvest surf clams with hydraulic clam dredge technology and to drag ocean scallops. Clearwater has the exclusive right to fish the entire Canadian quota of Porbeagle shark. Clearwater's May 1994 purchase of Sweeney's Fisheries in Yarmouth made that company the world's largest producer of scallops. As part of the deal, the Sweeney's herring operation in Yarmouth and East Pubnico was purchased by the Blades family of Clark's Harbour—the same group behind the Seafreez operation in Canso. Concentration of fish harvesting and fish processing capacity in the industry has increased and threatens to intensify with the current troubled times in the Atlantic groundfishery. Further accumulation of capital by the corporate offshore endangers the independence of small- and intermediate-scale plants whose operations may be further marginalized as raw product becomes more difficult to access. With fewer buyers, independent fishermen will likely have to contend with lower prices for their fish.

Industry concentration and rationalization of the commercial fisheries has long been supported by federal politicians and bureaucrats. A memo from federal Fisheries Minister, Jack Davis, to the Trudeau cabinet in 1970 states in part, "...Rationalization can lead to significant economy, fewer plants and fewer fishermen can lead to significant economies of scale." One of the last measures taken by the outgoing Conservative Government was to commission a study of the existing income structure in Canada's Atlantic fishery to determine a master plan for rationalization of the industry. Chaired by Richard Cashin, the report was received in November 1993, by the new Minister of Fisheries and Oceans, Brian Tobin. The Cashin Report reiterated old news—too many harvesters use too many boats with too much gear trying to supply too many processing plants by finding and catching too few fish. The report recommended the establishment of a Fishing Industry Renewal Board, which would implement capacity reduction, through the number of fishing licences available or the number of processors allowed to remain in business. Though proposed by the federal government, Renewal Boards have yet to be mandated to carry out industry rationalization.

The Cashin Report carried the same theme as the Kirby Report of December 1982, which was delivered by the Task Force on the Atlantic Fisheries. The Kirby Report admitted that saving the offshore industry and the jobs associated with it was of fundamental importance and justification for the bail-out. The bottom line was that all three—Davis, Cashin, and Kirby—were recommending the traditional inshore fishery, which used the innocuous hook and line, could be sacrificed as inefficient. With the inshore harvesters would go the independent fish processing sector—the backbone of the economy for many coastal communities.

Clearwater President, John Risley, seemed unable to conceal his enthusiasm for the Cashin Report and its recommendation for industry rationalization. Risley argued that fishermen must learn to stand on their own without assistance from Unemployment Insurance, and for those that remain in the processing sector, the gauge must be commercial viability. He suggested that smaller independent processors (largely family owned, employing twenty or fewer workers) should be the first ones to go, while large-scale enterprises with newer, more modern fish processing plants should be allowed to remain to form the core of Canada's east coast fishery. According to Risley, quoted at the Environment and the Economy Symposium of February 12, 1994, "The plants will need an option of capacity. Essentially, those people who have the poorest and most inefficient plants would surrender that plant processing capacity at the lowest possible price. Those who have more modern plants that were more recently built and expensive to build would obviously demand a higher price and they would be kept in the fishery as they should be because we obviously need that capacity to compete."

Of course, Risley's position is advantageous to Clearwater, which has attained status as the third largest groundfish operation on the coast behind National Sea and FPI, in addition to its position as the leading firm in the Atlantic shellfish industry. It is this company, above all others, that has been involved in government-assisted expansion over the past ten years, buying up old facilities or building new ones, largely on government credit. Clearwater is now in an ideal position to prey upon groundfish operations experiencing difficulty during the crisis, buying these enterprises and their holdings to further solidify its hold on the industry.

Industry rationalization is also the main thrust of the policy paper "Vision for the Atlantic Fisheries," presented by the Fisheries Council of Canada and released at their annual meeting in the fall of 1994. The big company vision for the Atlantic fishery calls on the federal government to dispense with its longstanding policy prohibiting the offshore sector and the inshore processing sector (presently licensed only to process fish) from acquiring harvesting rights to the inshore portion of groundfish quotas. Should the federal government oblige, the move would allow the offshore companies to increase their share of the groundfish harvest and allow inshore processors the opportunity to better control supply and price of the raw product. Inshore fishermen are vehemently opposed to what would be a fundamental change in policy on harvesting rights—one which they argue would mean either having to sell their operations or fish under contract for a processing company.

Both the Cashin Report and the "Vision for the Atlantic Fisheries" have generally received poor reviews. Inshore fishermen have made it clear that they see these proposals as part of a larger plot to wipe out their sector, leaving the large corporations in control. The Cashin Report also elicited a negative review from the Canadian Oceans Caucus discussed on CBC "Radio Noon" on December 21, 1993. Irene Novaczek, Chair of this activist group, pointed out that the report was geared largely to the Newfoundland reality, although communities in the rest of Atlantic Canada were in crisis as well. She said it was far too early to say that 50 per cent of the inshore sector had to lose their groundfish licences as the Cashin Report implies. This inshore sector is overwhelmingly represented by hook and line fishermen who pose little or no danger to fish stocks.

Novaczek emphasized that the public has to be aware that the groundfish crisis was primarily caused by allowing destructive fishing power to ride roughshod over the marine environment, and future success depends on recognition of this fact. She remarked, "There is certainly not enough emphasis in this report on the absolute necessity to consider the limitations of the ecosystem in trying to devise any future plans for the fishery." Expounding on the virtues of a hook and line fishery and cautioning against destructive fishing practices, she went on to suggest that a fundamental examination should be conducted on the costs and benefits offered "in both ecological and broader social terms" by the corporate offshore sector which employs large trawlers. Novaczek suggested that an in-depth analysis of the types of technology used in the fishery and an examination of the DFO's fisheries management system would show the inshore fixed gear fishery to be sustainable and would call into question the continued use of dragger and trawler technology.

It is obvious that an analysis of the appropriate way to harvest our groundfish resource should have been conducted by now. Our scientific community should know which fishing methods and technologies are a threat to the sustainability of the groundfishery. Management systems should have encouraged the use of non-destructive technologies and harvesting practices, and fishery managers should have worked to ensure conservative levels of allowable catch so that our groundfish resource was not over-exploited.

It is true that the extension of Canadian jurisdiction in 1977 and the subsequent need to restore groundfish stock health placed a tremendous amount of pressure on scientists employed by DFO. The established priority for scientists was measurement of the groundfish biomass and determination of an acceptable level of harvest. Exploring the fundamental questions of the suitability of the harvesting technology was not a priority. Instead, biologists were entrusted with determining the year-to-year status of each of about thirty-five Atlantic groundfish stocks and were required to make reliable predictions on the future growth of the stocks based on the raw data collected. The complexity of marine biological systems, such as predator-prey relationships, and the interplay of changes in the physical environment, such as water temperature, made it virtually impossible to conclusively analyze the data. Recommendations for management were, of necessity, hedged as "optimum levels of exploitation" with the acknowledgement that these must be set and later adjusted by a process of trial and error. Further, it was recognized that natural variability within the marine system would continue to play a significant role in determining stock health. These environmental factors could obscure the effects of resource overexploitation should industry take more than the allotted quota, making it increasingly difficult to institute the management measures needed for a sustainable fishery.

Into the mix of environmental factors and harvest rates to be considered by scientists were stirred heaps of enthusiasm and high expectations that swept the groundfish industry in 1977 with extension of Canadian jurisdiction. The industry was optimistic that the domestic harvest would soar with the expulsion of the foreign fleets; this expectation was a factor considered by some scientists when making their predictions on stock health. Divergent forecasts were being floated by scientists in the early 1980s. One forecast indicated that groundfish stocks, in particular northern cod stocks, were facing decline while another indicated stocks were increasing dramatically.

Managers and politicians, eager to hear an optimistic forecast, naturally subscribed to the promising reports and set northern cod total allowable catches at the upper end of the range. Added incentive for a higher TAC came from the offshore sector's encouragement. According to original conditions for the split in northern cod TAC between the inshore and offshore sector, the inshore had been guaranteed that their portion of the TAC at 115,000 tonnes per year would not be touched, unless the total TAC for both sectors was reduced below 115,000 tonnes. Any reduction in the TAC for northern cod would, therefore, have to come from the offshore portion. This led to immense pressure from the offshore lobby, which did not want to see the northern cod TAC—and their share of the catch—reduced. Despite the re-evaluation of stock health by CAFSAC scientists in 1988 and the Harris Report which found that northern cod were in trouble, a series of fisheries ministers, including Tom Siddon, Bernard Valcourt, and John Crosbie, overruled the call of scientists to cut the northern cod TAC by half, opting to delay the inevitable closure of the fishery.

The scientific community had been coopted by its political leaders. Senior bureaucrats and politicians did not want to hear the bad news, so those scientists who would sound the alarm bells were given no audience. Even scientists with tenure were reluctant to speak out because their careers were dependent upon government funded research grants. We can rely on scientists to recognize the signs of stock decline when provided with accurate information, but because of the environment in which they work, we cannot expect them to go public with their concerns. It is clear that the role of fisheries biologists is simply to advise their superiors. To bring attention to inappropriate conclusions drawn from groundfish stock assessments by senior bureaucrats and politicians may result in career penalties or other forms of discipline.

Acquiescence is inherent in the system. This is not to accuse fishery biologists and oceanographers of ulterior motives or to imply naivety. Over one-half of all scientists around the world work either for military establishments or for private industry. Their purpose is either to make weapons or to design goods for profit. The portion that work for government most often find themselves cloistered away in research institutes and limited by the agenda of their political sponsors.

David Suzuki, television and radio host, geneticist, and environmental activist, was a speaker addressing the Environment and the Economy Symposium at Dalhousie University in February 1994 on the crisis in the Atlantic coast fishery. He shared his concerns for the integrity of the scientific process. "...I know for a fact that they are frustrated and demoralized because they see their scientific research being massaged to fit political agenda.... And I know that there are DFO scientists who have known for years that there is a crisis in the groundfishery but who have been frustrated by the political structure within which they exist...."

Suzuki could have been referring to Jeffrey Hutchings and Ransom Myers, two DFO biologists who dared to confront the official line maintained by their superiors that northern cod stocks had been the victim of cold ocean temperatures. These two scientists explored the possibility that northern cod stocks were overfished by Canadian fishermen after DFO had set the total allowable catch unrealistically high. They then allowed their findings to be published in the April 1995 edition of the American science magazine *Discover*. Hutchings and Myers analyzed research data on cod stocks dating back to 1946. The historical data showed that by 1977 the number of spawning stock off Newfoundland, critical to the health of the species, had declined by 94 per cent from that in 1962. The stock was in dismal shape, and the Canadian fishing industry should have been instructed to slow down to allow the spawning stock to recover. The poor health of the stock coupled with overly optimistic predictions of cod stock growth made by DFO biologists at the time set the stage for disaster. Hutchings and Myers suggest the damage from overfishing could have been reduced had politicians cut the northern cod quota by half as recommended by CAFSAC biologists in 1989.

According to *Dalhousie News* of April 26, 1995, since the publication of their findings the two scientists have had to endure harassment and "lots of repercussions" from government decision makers. Nevertheless, they take solace in the fact that their findings were made public, and they have been encouraged by fishermen and scientists who concur with their findings.

Robert Fournier shared the stage with Suzuki at the February 1994 Environment and the Economy Symposium. Fournier is a Professor of Oceanography and the Associate Vice-President of Research at Dalhousie University. He also served as a member of the Northern Cod Review Panel chaired by Harris. Fournier spoke of an exciting relationship that was established in 1990, when three fish companies—National Sea Products, Fishery Products International, and Clearwater Fine Foods—joined forces with the federal government in the creation of the Ocean Production Enhancement Network (OPEN). Their mission was to delve into the factors contributing to the health and welfare of the Atlantic cod and the sea scallop, and to work to enhance the commercial production of the ocean. Each of the offshore giants invested $1 million while federal Fisheries Minister, Bernard Valcourt, chipped in $20 million of taxpayers' money.

OPEN, set up under the auspices of the Network of Centres of Excellence, reaches nation-wide, embracing the work of researchers at eight universities and three government laboratories. The OPEN program has an impressive list of 250 funded scientists, including geneticists, molecular biologists, and other research specialists. The work conducted thus far has focused on stock differentiation, cross-shelf migration, recruitment, temperature tolerance, and predator-prey relationships. All of these fields of inquiry are relevant to an understanding of some of the issues surrounding stock health. Strangely, there appears to be little interest in the effects of overfishing or destructive fishing practices, such as groundfish and scallop dragging. Effort has been directed toward finding an environmental cause for groundfish stock decline and for ways to enhance production of commercial stocks. From a distance, it appears that the agenda of OPEN is directed by the corporate offshore presence—directed away from research which may implicate the large companies and their fishing methods, and directed toward research which may provide them with a competitive advantage. In all this, the scientific community, desperate for research funding, is forced to comply with the direction given.

An agenda determined by corporate subscribers can decide the outcome of research. Prior to World War II, a debate focused on the fishing of herring stocks off the coast of California. The California Fish and Game Commission, whose primary role was to protect the state's marine resources, had cautioned that commercial prosecution of the fishery could not exceed certain limits without leading to stock collapse. Federal scientists working for the United States Bureau of Commercial Fisheries were assigned the task of working toward ocean production enhancement for the American fishing industry, so they looked for causes other than overfishing for the decline. The result was that each group of scientists, using the same data base came to different conclusions which were in line with the agencies for which they worked.

John Radovich, reviewing the history of the sardine collapse in *Lessons from Coastal Upwelling Fisheries,* noted the ideological struggle waged by scientists from each camp: "Scientists are directly or indirectly influenced by the values of their society, their institutions, their academic disciplines, as well as by their personal political beliefs. Each scientific discipline is saturated with values imposed by its specific profession, and scientists are influenced by the agencies for which they work and to which they owe some allegiance. Thus, the definition of a problem becomes a biological one, a physical one, an economic one, a psychological one, a sociological one, even a philosophical one, depending on the researcher's discipline." Of course, the answer to the problem may differ, depending on the researcher's field of study and the value system under which he or she works.

The Radovich review noted that scientists from the Scripps Institute of Oceanography in California were also attracted to the debate about the sardine collapse and formulated their own theory of the collapse, which leaned toward the investigation of "large-scale processes" (current, temperature, and other physical and chemical processes), rather than the impact of fishing effort. In other words, oceanographers at the Scripps Institute sought to explain the sardine collapse strictly in oceanographic terms, precluding any serious consideration of the impact of fishing effort. It is noteworthy that oceanography was chosen by the federal government and offshore companies as the lead discipline under the Ocean Production Enhancement Network, thus predetermining the focus of studies carried out under the program.

Oceanographers are accustomed to assuming the lead role in multidisciplinary studies. Radovich explains the hierarchy in *Lessons from Coastal Upwelling Fisheries:* "...Oceanographers frequently define their field as encompassing the ocean and all the sciences that are studied in relation to the ocean. This all-inclusive perspective relegates other sciences (biology, chemistry, physics and geology) to the position of sub-disciplines of oceanography. Such a disciplinary perspective tends to focus attention away from the effects of local human activities on various marine resources and to extend efforts instead toward the investigation of large-scale processes in search of fundamental generalizations to explain widespread phenomena. One might argue that elitism tends to develop, for example, at the top of the scale the physical oceanographer, and at the bottom, the biological oceanographer. From the viewpoint of a school of oceanography, the solution to most fisheries problems invariably involves a large-scale, multi-vesselled, physical and chemical assault on a large part of the world ocean, because that is how the problem is conceived—by definition, of course."

In other words, an oceanographer's answer to the depletion of groundfish stocks would not, by definition, include the examination of destructive fishing practices. Likewise, it would not involve studies to determine the proper adjustment of mesh sizes employed by a trawler's codend to reduce the take of small fish. Even when such studies were authorized and conducted by DFO scientists, there was a good chance that their recommendations would either be ignored or overridden. The corporate lobby, concerned with dispelling any impediment that might stand in the way of full resource exploitation, could always fall back on its firmly established alliance with government to remove that obstacle.

Such was the case with one of the key recommendations made by the Scotia-Fundy Groundfish Task Force, established to recommend improvements to fishery management. The Hache Task Force recommended that a larger minimum mesh size be made mandatory for the mobile gear sector. While the recommendation was made as a conservation measure that would allow smaller fish to escape the net and spawn later, the measure was equally motivated by the need of fishery managers to control the pressure exerted on stocks by the overcapacity in the inshore fleet.

The standard diamond mesh used on the codend of a standard trawl tends to close when the weight of the catch strains the gear. The central meshes ahead of the codend tend to elongate and close, making it difficult for even the smallest fish to escape. With a properly rigged square mesh codend, the meshes are expected to stay open, even with a load of fish. Experiments conducted for DFO showed that 140 mm square mesh and its equivalent, 155 mm diamond mesh, were the ideal size to allow for release of fish less than 43 cm (17 in.), the legal catch size.

The new mesh size for codends of standard trawls, to be in place by March 1, 1991, called for a minimum of 140 mm square mesh or 155 mm diamond mesh. The change from the previous minimum of 130 mm diamond mesh applied to draggers and trawlers fishing Scotia-Fundy waters. It was reasoned that with the concurrent increase in minimum fish size from 41 cm to 43 cm (16 in. to 17 in.), the larger mesh size would reduce the potential for discards since the smaller fish would be able to escape. There would also be less need to dump fish to meet trip limits on certain species because the reduced caches per tow would allow more control over the amount of fish brought aboard. This control would thus allow the captain to edge closer to the trip limit and reduce the chances that the trip limit would be exceeded by landing a large catch on the final tow. Haddock stocks were, and still are, overexploited, and the mandatory increase in mesh size was forecast as potentially very important for that stock because haddock tend to run smaller at age than cod or pollock. The increase in mesh size was predicted to reduce trawler efficiency for haddock more than for cod and pollock, thus serving as a conservation measure which would permit more haddock to reach spawning maturity.

The increased minimum mesh size would mean that the number of draggers and trawlers engaged in the Scotia-Fundy groundfishery would not be reduced, but they would have to fish longer to catch the same amount, albeit a larger, more valuable fish. This would reduce the effective capacity of the fleet. The inshore dragger fleets preferred the new mesh requirements over more Draconian measures to reduce fishing capacity, such as loss of quota or reduction in fleet size. National Sea Products, on the other hand, cried undue hardship, claiming classified studies they had conducted indicated it would cost its offshore fleet an additional $10 million, annually, to catch the same amount of fish. The company lobbied long and hard, and to the chagrin of the inshore fleet, the federal government and Fisheries Minister John Crosbie reneged on the new mesh sizes in July 1991. Crosbie concurred with National Sea. Although his department had not seen evidence of such costs from catches it had analyzed, according to *The Chronicle-Herald* on July 4, 1991, the costs to National Sea of fishing with the new mesh size were considered too great to ignore. Rather than put additional strain on the big companies, he sacrificed conservation.

After only four months in effect, the new regulations were canned in favour of a compromise which allowed for 130 mm square mesh and 145 mm. diamond mesh. Interestingly, the other offshore giant, Fishery Products International, had been using square mesh exclusively and with no formal complaints. M. J. Hogan, director of harvesting research and development for FPI, was quoted in *The Chronicle-Herald* on June 21, 1990, to say the

Newfoundland company's offshore trawlers had been using square mesh exclusively in their cod fishery "with quite a bit of success."

The DFO's explanation of why it was appropriate to return to the 130 mm square mesh minimum, was difficult to pin down, with its conflicting figures on the percentage of small fish retained by the smaller mesh. Earlier, in a press release of May 10, 1990, the DFO revealed the results of a study which showed that 130 mm codends retained between 25 per cent and 35 per cent of fish under 43 cm (17 in.), while tests conducted on 140 mm square mesh indicated a retention rate of undersized fish of only 4 per cent to 6 per cent within the target range sought by researchers. According to statements made in July 1991 by John Angel, Regional Director of Fisheries and Habitat and lead official in charge of implementation of the recommendations of the Hache Task Force, (just prior to Crosbie's reversal) the 130 mm square mesh or 145 mm diamond mesh would increase the undersized fish caught by only 6 per cent to a total of 11 per cent. Some in the industry claimed that the estimates were low. Perhaps they had remembered the DFO press release of fourteen months earlier.

With the return to smaller mesh sizes, it was obvious that even the best advice made public could be overridden by the lobbying power of the offshore. For some this was an awakening, for others it was a reaffirmation. There were a series of issues in which the government of the day had abandoned its responsibility to resource conservation and yielded to the large corporations—government's refusal to cut northern cod quotas following the advice of CAFSAC scientists in 1989; the refusal to investigate the impact of dragger technology on groundfish habitat; government licensing of a factory freezer trawler; granting an excessive portion of groundfish quotas to the offshore under the EA system; and the list goes on. With the reversal on the mesh size, fishery managers and the Minister of DFO showed they were willing to try to cover up their in-house findings simply to accommodate big business. The integrity of the scientific and management process had been breached in the name of the corporate bottom line.

Meanwhile, the emphasis on oceanography at Dalhousie and Memorial universities and the working relationship of these two institutions with the Bedford Institute of Oceanography in Dartmouth and the Northwest Atlantic Fisheries Centre in St. John's has meant the channelling of most of the research effort into environmental causes for the groundfishery decline. This preoccupation with environmental causes, nurtured by research funding, led to the staging of a workshop on northern cod in Newfoundland in January 1993. It was followed by the Symposium on Cod and Environmental Change, held in February 1994, at the Bedford Institute of Oceanography. Each of these gatherings included a series of presentations from internationally renowned researchers stating their case for environmental cause, such as temperature, wind, and salinity which directly impacted on the cod, and the play of these same factors upon the growth of plankton, the food supply of newly hatched cod. There was virtually no mention of overfishing as an important factor in the demise of groundfish stocks. As of February 1994, nearly two years after a moratorium was imposed on northern cod fishing and a year after much of the remainder of the region was closed to fishing, the government was still unwilling to admit that overfishing was a primary cause of the groundfish crisis in Atlantic Canada.

Furthermore, the 1994 report of the National Advisory Board on Science and Technology titled "Opportunities from our Oceans" recommended that marine environmental research in Canada be placed under the auspices of resource development and management. Making marine environmental research subservient to the resource development and management agenda is cause for concern that the new direction will only expedite resource exploitation and ocean production enhancement—the large company corporate agenda.

The takeover of research effort by those in charge of a corporate agenda is not unique to North America. The Australian oceans policy has likewise suffered from lack of a comprehensive overview and integrated action. The 1993 McKinnon Report on Marine Sciences and Technology has recommended that the government 'down under' establish a national marine science and technology council. Although the Australian government has announced it will adopt this suggestion, there is still some uncertainty about its composition and terms of reference. It is likely that emphasis will be placed on links with industry with regard to decision making and project funding and on short-term research and development payoffs, as in Canada, where OPEN is committed to research to enhance production of cod and sea scallops, two key commercial species. A wider vision of commitment to resource sustainability appears to be receiving short shrift in Australia as it has in Canada. The Australians are fearful that the short-term agenda of the corporations will dominate. The Australian government is demanding that at least 30 per cent of the funding for research come from the private sector and as a result, they are willing to sacrifice a shift in emphasis from basic research to applied research, in other words, from the study of how living things and their surroundings interact to a study of how to enhance the production of commercial stocks.

A closer examination of Canada's Ocean Production Enhancement Network reveals a roster laden with well-connected public and corporate players—a "who's who" of government, big business, and the scientific community. The Board of Directors of OPEN is chaired by Robert Fournier. The twelve-member board includes representatives from each of three offshore companies, as well as the formal lobby organization for large fish companies, the Fisheries Council of Canada. The spring 1994 line-up included the replacement of Stephen Greene as Clearwater's appointment on the

Board with Eric Roe, the company's Director of Public Relations. Greene was Assistant to the President, but on February 1, 1994, he took up his new position as Chief of Staff for the Reform Party in Ottawa. It is also interesting to note that Sam Naidu, Clearwater's appointment to the Scientific Management Committee, was being replaced so that he could resume his role as a federal government employee with the Pelagic Fish, Shellfish, and Marine Mammals Division, Science Branch of the DFO in St. John's, Newfoundland. It is at least unsettling to realize that some in the scientific community float between employment with the DFO and employment with private industry. If the foundation of science is objectivity, its foundation is shaken.

Roger Stirling, President of the Seafood Producers of Nova Scotia, also represents the Fisheries Council of Canada. It is interesting to note that William Robertson, Manager of Canadian Fish Operations for Connors Brothers Limited of Black's Harbour, New Brunswick—a major player in the east coast herring and aquaculture industry—is Chair of the Industrial Advisory Board (IAB). Brian Giroux, advisor for the Atlantic Coast Scallop Fishermen's Association serves in that capacity on the IAB, though he is more widely recognized as Executive Director of the Nova Scotia Dragger Fishermen's Association. Ginette Robert, a scallop scientist with DFO in Halifax, represents the Nova Scotia Offshore (Scallop) Licence Holders Joint Venture on the Industrial Advisory Board.

Could we really expect anyone of this group to support funding for research on the effects of destructive fishing practices, when the interests they represent employ the same methods in the commercial pursuit of groundfish, scallops, and clams? Scientists associated with and funded by OPEN would surely realize that their scientific inquiries should be designed to assist in production enhancement rather than ecosystem protection. Research into ocean production enhancement is to be encouraged, but promotion and funding for such research should be left entirely in the hands of industry and the private sector. On the other hand, ecosystem protection research, which has to this point been hampered by lack of government support and funding, should be freed from the tentacles of corporate influence. To achieve this, government would have to divorce itself from research partnerships with big business. Through the DFO, universities, and independent research institutes, government should adequately support and fund basic research into ecosystem protection and allow the scientists involved in such projects to maintain objectivity.

Instead, the corporate influence presently reaches Dalhousie, Laval, McGill, Memorial, Simon Fraser universities, the Universities of New Brunswick and of Quebec at Rimouski. Not only are lead scientists at these institutions funded by OPEN but also an entire generation of graduate and undergraduate assistants working toward their degrees. All are exposed to the corporate agenda of ocean production enhancement.

It appears that corporate influence will continue under the new Liberal administration. Federal Fisheries Minister Brian Tobin and Minister David Dingwall, in charge of the Atlantic Canada Opportunities Association (ACOA), have announced funding of $3 million for 1994-95 to continue the work of OPEN. Oblivious to the danger inherent in the continuation of the program, both ministers expound the virtues of OPEN, according to the *Sou'Wester* of April 15, 1994. All the right words, with all the wrong reasons. "Ongoing collaboration between industry, universities and government scientists is essential to improve our advice to fisheries managers and to help restore this resource," said Tobin. "We value this contribution and are pleased to provide support."

Dingwall commented in a similar vein, "New partnerships that cut across regional and disciplinary interests are a key to meeting the economic challenges of the future.... This kind of combined action will contribute to a revitalized fisheries economy and create opportunities for the high-technology oceans industry to provide competitive goods and services."

In an attempt to revitalize the sagging economy of Atlantic Canada and to restore groundfish stocks to health, the federal government has chosen to hitch its wagon to the wily promise of development of a high-technology oceans industry offered by the self-interested corporate offshore. Ottawa has determined that it will update its long-time alliance with big business by pursuing a new age partnership. This new partnership necessitates the acquiescence of the scientific community and promotes ocean production enhancement, rather than dealing with the primary reasons for the crisis in the groundfishery—overfishing and the use of destructive fishing technology. As well, government has indicated its willingness to join hands with the corporate offshore in a move to "rationalize the groundfishery," a phrase that means reduction in the size and significance of the competing inshore harvesting and processing sector. Big business has always wanted the entire fishery to itself and is more eager than ever to seize upon the opportunity, should government comply. Sadly, it appears that the people's elected representatives are willing to be a part of the takeover by the offshore corporations of the Atlantic coast fishing industry.

Chapter Eight

Quotas and the Food Chain: Will Partial Restrictions Work?

Though the crisis in the Atlantic groundfishery was predicted, it was also avoidable. For years, fishermen and industry watchers had warned the build-up in fishing capacity and the shift to more powerful harvest technologies would eventually cause groundfish stocks to collapse. They forewarned government that overfishing would decimate stocks. They warned that dragger and trawler technology was damaging the habitat of groundfish stocks. They pointed out that groundfish must eat too, and a government which allowed that foreign and domestic interests to hammer away at food stocks for groundfish was flirting with catastrophe. They sounded all sorts of alarms that went unheeded.

From the fourteenth century, inshore and small-scale fishermen have warned of the inevitable destruction of stocks through a build-up of harvesting capacity. They cautioned that more sophisticated gear and fishing techniques would put undue pressure on stocks, altering the balance beneath the waves. Complaints were registered about the use of trawl gear from its inception. According to Sahrhage and Lundbeck's *A History of Fishing,* complaints were voiced since the thirteenth and fourteenth centuries by British inshore fishermen working the Thames River about trawlers using small meshed nets, taking undersized fish, and destroying bottom life and habitat on which more commercially valuable species were dependent. They also pointed out early on that in Britain's unregulated fishery, trawlers were encroaching on their traditional grounds and stealing markets from under their noses.

Similarly, Canadian fishermen were quick to lodge complaints when the first trawlers were introduced on this side of the Atlantic at the close of the nineteenth century. Although this later version of the trawler was improved over its earlier relative, Canadian opposition was well-organized and succeeded in preventing a build-up of trawler fishing power until after World War II.

Thereafter, the government succumbed to pressure to modernize the fishery and increase production.

The worst fears of fishermen became reality with the explosion of fishing power after World War II. The growth of applied science brought new technology for navigation and fishfinding, and it brought improvements in fish handling. The heavy overfishing of the northwest Atlantic by the distant water trawler fleets through the 1960s and 1970s confirmed the fears expressed earlier by inshore fishermen. It became possible to devastate the resources of a sea as vast as the northwest Atlantic.

The Canadian government had little power to control the pillage inflicted by the foreign fleets which, prior to 1964, fished waters to within 5 km (3 mi.) of the coast. In 1964, Canada took unilateral action to declare control of waters within 14 km (9 mi.) from shore, after years of negotiations with the international community had failed to produce an agreement. Although the International Commission for the Northwest Atlantic Fisheries (ICNAF) had been established in 1949 to undertake research into groundfish stocks and bring order to the fishery, it was ineffective and the groundfish stocks crashed in the late 1960s.

Extension of Canadian territory to 200 miles from shore provided a temporary respite for groundfish stocks. Northern cod stocks off Newfoundland and Labrador increased rather dramatically from a Canadian catch of 103,000 tonnes in 1978 to a catch of 214,000 tonnes in 1983. The total Atlantic groundfish catch jumped from 535,000 tonnes in 1978 to 728,000 tonnes in 1983. However, the resource was not being shared equally between the inshore and offshore sectors. The Newfoundland and Labrador inshore sector experienced some particularly poor seasons, while the offshore trawlers were filling their nets and enjoying increasing catches. Having used cod traps from the same berths or hook and line on the same inshore grounds for generations, these fishermen were at first perplexed by the failing catches. While they cautioned that something was terribly wrong, the DFO offered only platitudes.

The official explanation from DFO was that inshore catches may have been suffering from unusually cold water temperatures or catches were down because the cod decided to stay offshore to chase capelin and other sources of food. To inshore fishermen, it made more sense that fish, which had traditionally migrated to the inshore grounds with the yearly warming of coastal waters, were being intercepted by the powerful mobile trawlers. While inshore fishermen made bold accusations that trawlers and draggers were at fault, government failed to take seriously their recommendations. The Newfoundland Inshore Fishermen's Association (NIFA) commissioned its own study into the decline of northern cod stocks in 1986. Using data provided by fishermen and DFO, the Keats Report, compiled by Keats et al., predicted the certain collapse of the stock, but their call for drastic cuts to the northern cod total allowable catches was ignored. Inshore fishermen were instead treated by federal authorities as unreliable sources of information; their opinions were disputed and their claims were disallowed. Yet, inshore fishermen are the true scientists. Their livelihoods depend on an expert knowledge of the habits and tendencies of groundfish; this accumulated knowledge—including details of the locations and size of catches, the amount and type of gear used, and the conditions under which the catches were made—is both reliable and objective. Had their expert opinion been heeded, the crisis could have been avoided.

As it was, fisheries managers gave credence to the information supplied by the offshore, believing that the highly organized offshore companies, with their sophisticated record keeping, could provide more reliable data inputs on which to base stock health assessments. It was much easier to contact the head offices of FPI and National Sea than to compile the information from thousands of inshore fishermen. However, fishery managers failed to take into account that the offshore trawler fleet was exerting more effort to return the same catch, which is in itself an indication of failing stock health. In addition, DFO scientists charged with advising on the health of groundfish stocks, failed to account for the tremendous amounts of fish being discarded and dumped by the trawlers. It was not until the mid-1980s, with groundfish stocks showing visible signs of decline that these destructive fishing practices were taken seriously, according to Angel et al.

The effect of destructive fishing practices employed by the company trawlers not only imperilled stock health, but also threatened to jeopardize the portion of the total allowable catch available to the offshore sector. Offshore companies such as National Sea and Fishery Products International had a vested interest in sustained total allowable catches because under the Enterprise Allocation program, established in 1982, government had guaranteed the inshore sector a yearly quota of 115,000 tonnes of northern cod. Therefore, the offshore companies were first in line to suffer quota losses should the total allowable catch of the all-important northern cod require reduction. It was therefore in the best short-term interests of the offshore to portray a healthy cod stock and oppose any suggestion of a reduction in TAC for northern cod. It was the ability of the offshore to lobby government and the catch report they presented to DFO suggesting cod stocks were not in trouble that kept the TAC at a higher than justified level for a number of years. As a result, northern cod stocks were fished down to unsustainable levels.

The dismissal by DFO of the inshore position that quotas had to be cut and law and order brought to the fishery was no easy feat. Inshore fishermen admitted to the illegal practices of misreporting, underreporting, discarding, and dumping at the same time as they accused offshore fishermen. Inshore fishermen warned that the statistical inputs used by fisheries biologists were often the product of efforts by fishermen in all sectors, including their own, to mislead enforcement officials and avoid prosecution. They justified their transgressions as necessary when competing with others who exercised much less restraint. For many, fishing had become a ruthless struggle to survive, but it was the groundfish stocks that were suffering the most. Requests by fishermen to stop the abuse by cutting quotas was a call for help.

Along with the alarming decline in the number of groundfish frequenting the inshore grounds, veteran fishermen became concerned about decreases in the average size of individual fish. They argued this decrease in size was also the result of excessive fishing pressure. Fewer and fewer of the large and more fecund female fish were being landed in the 1960s and 1970s, which should have served as a warning sign to fishery biologists concerned about the ability of the stock to replenish. A minimum catch size of 41 cm (16 in.) was in effect but seldom enforced. Inshore fishermen fishing the Scotian Shelf suggested that the minimum catch size be increased to 48 cm (19 in.). This minimum for groundfish was advanced during the mid-1980s to deal with the pending crisis in the Atlantic groundfishery, though the concept of allowing fish to reach sexual maturity and allowing them the opportunity to spawn a number of times before being harvested was a time-honoured precept. As well, the larger fish would attract a high market price.

An increase of legal catch size received considerable support among those making suggestions to the Hache Task Force on the Scotia-Fundy groundfishery. The end result was very disappointing. The Task Force felt that moving to a minimum fish size of 43 cm (17 in.) was sufficient hardship because the industry would require a larger mesh size for dragger and trawler nets, more time would be required and more effort would be expended by the mobile gear sector to catch the same weight, though fewer fish would be harvested.

At the unveiling of the Hache Report in January 1990, industry representatives in support of the 48 cm (19 in.) minimum catch size were critical of an increase of only 2.54 cm (1 in.) as "too cautious" and "a recipe for disaster," suggesting it did not go far enough to protect immature fish. Failure to impose a larger minimum size, they said, would endanger the chance for recovery of the fishery. "You'll never rebuild the stocks when you kill all the babies and don't give them a chance to reproduce," Basil Blades, owner of Sable Fish Packers of Cape Sable Island, told the chair of the Hache Task Force. He said his plant had recently received 49,830 kg (110,000 lbs) of fish trucked from Cape Breton and reports from his foreman indicated none of the fish measured 48 cm (19 in.) and very few reached 43 cm (17 in.). *The Guardian Newspaper* of January 9, 1990, quoted Blades: "Instead of going to a larger net to produce larger fish, the Task Force chose to go to a net that produces 17 in. fish. I would like to suggest to you that a 17 in. fish has not reached sexual maturity, has not reproduced even once and when you look at what was on my doorstep yesterday morning, there was 110,000 lbs of fish that had never reproduced."

Noble Smith of the Sou'West Nova Longliners Association also expressed concern for the smaller minimum fish size and explained that plant workers realize the implications for stock health. He was quoted in *The Guardian Newspaper* on January 9, 1990: "I asked some plant workers if they had any work. They said they had work coming up tomorrow. 'We're working on a truck load of abortions!' They are fed up with working on small fish. They are calling them abortions."

Newfoundland inshore fishermen were particularly concerned about fishing spawning aggregations. They argued that towing a net through an aggregation of spawning fish disrupts the mating process, placing males and females under undue stress and jeopardizing reproductive success. This was the case formally made by the Newfoundland Inshore Fishermen's Association (NIFA). In 1989, the association launched a suit against the federal government alleging that Ottawa had mismanaged the stocks by failing to reduce quotas to reflect declining stock health and by allowing the offshore to target vulnerable spawning aggregations. President of NIFA, Cabot Martin, asked the court to issue an injunction to stop trawlers from fishing the northern cod spawning grounds during the spawning season. The group also asked for an environmental review of the practice of setting quotas and establishing access to the resource.

Thanks to delays best attributed to the work of the strong offshore lobby, the answer to the NIFA's primary concern was granted three years later by Fisheries Minister John Crosbie. By this time irreparable damage to the northern cod stock had been done. Faced with yet another discouraging report from the Canadian Atlantic Fisheries Advisory Committee (CAFSAC), Crosbie announced on February 24, 1992, that, along with other measures, such as cutting northern cod quotas for the first half of the year and closing much of the offshore grounds, trawlers would not be permitted to harvest offshore concentrations of northern cod during peak spawning periods. These measures were overriden five months later, when a total ban on the fishing of northern cod was imposed. The recommendation to reduce trawling of northern cod at peak spawning period was too late; the chance for recovery was lost.

Still smarting from the closure of the harp seal hunt largely forced by animal welfare groups, inshore fishermen in Newfoundland actively pursued the support of environmentalists in their lawsuit against Ottawa. Farley Mowat, outspoken critic of the hunt of marine mammals, was enlisted in 1989 in preparation for NIFA's court battle with the federal government. The sage environmentalist cut directly to the quick, calling on federal fisheries managers to respect the legitimate concerns of inshore fishermen and stop paying court to the masters of industry. "Destructive practices such as trawler fishing on spawning grounds during the spawning season must cease," he wrote. In the same letter to NIFA, printed in Martin's *No Fish and Our Lives,* Mowat promised to carry the concerns of Newfoundland inshore fishermen with him across Canada and around the world. "I will be pleased to tell all I meet of your fight to protect the northern cod stocks and Newfoundland's outport way of life."

Warnings of the effects of disruptions by trawlers to fish mating were cryptic in the Harris Report of 1990. Following a series of public exchanges with the Review Panel, Newfoundland inshore fishermen were confident that the final report of the "Independent Review of the State of the Northern Cod Stock" would come down hard on destructive fishing practices conducted by the mobile gear fleet; an official condemnation would be forthcoming. However, the final report was not so independent of big company influence. The issue of destructive fishing technologies and method was soft-pedalled and the recommendation to investigate the impact of fishing spawning populations did not reveal the panel's concerns beyond the effects on fish mortality and biomass reduction.

Inshore fishermen in Nova Scotia publicly accused the Department of Fisheries and Oceans of "gross mismanagement" of the fishery. Following on the heels of action taken by NIFA, fishermen in Nova Scotia took the federal government to court in 1993. The fishermen, organized under the Inshore Fishermen's Bonafide Defence Fund, claimed that DFO was part of a master scheme to push inshore fishermen out of the fishery and then turn the industry over to the large companies. They claimed that government should have heeded the early warnings. Had decisive action been taken by the mid-1980s, stocks would not have crashed and a subsistence fishery could have been maintained for the inshore until stocks were replenished. The northern cod moratorium in 1992 was a scenario the Inshore Fishermen's Bonafide Defence Fund claimed was part of this master plan, for only corporate interests would have the financial reserves and corporate flexibility required to ride out the closure. According to the association, quota cuts made in December 1992, hit the inshore harder than their offshore counterparts and were the result of years of DFO mismanagement, which gave no consideration to the future of the many coastal communities dependent upon the inshore fishery. *The Chronicle-Herald* of April 15, 1993, quoted Joe Boudreau, Chair of the association: "Everything the federal government does is for the offshore (fishery), plain and simple. We just cannot stand for it anymore."

The legal action was dropped in January 1994. The Nova Scotia Supreme Court ruled that the membership list of the Inshore Fishermen's Bonafide Defence Fund did not contain a significant cross section of the inshore sector and therefore did not sufficiently represent it. The membership consisted of healthy representations from Prince Edward Island, Cape Breton, and northern and southwestern Nova Scotia, but attracted only one member from New Brunswick and none from either Quebec or Newfoundland. The action was effectively stymied by the court's decision, though the Chair, Joe Boudreau, suggested the fishermen would continue with their cause by publishing the information they had gathered to support their case.

Fishermen in tune with the delicate ecological balance beneath the waves also appreciate the importance of feed stocks such as squid, shrimp, mackerel, herring, and silver hake to the health of commercial groundfish species. This appreciation is especially strong among hook and line fishermen who use some of these same species as bait aboard their vessels. They have come to realize that, when dealing with nature, we cannot fish one species hard without having an impact on those above and below it in the food chain. In *The Coast Guard* of February 1, 1994, Evan Walters of Barrington, Nova Scotia, a former fisherman and newspaper columnist, remarks on this truism, bemoaning the overfishing of herring stocks vital to groundfish stock health. "What we see is starving fish, starving birds, a few fish full of scrambling jacks, brittle stars and garbage. It appears that we have destroyed the local fisheries simply by removing the basic necessity of their lives, food!"

This awareness is shared by veteran fisherman Blandford Nickerson of Port L'Herbert, Nova Scotia. He is quoted in the *Sou'Wester* on July 1, 1995, addressing the House of Commons Standing Committee of Fisheries and Oceans at its stop in Barrington in June. Nickerson expressed his disappointment in the lack of recognition for the primary reason behind the crisis in the groundfishery—that there "is no food left in the ocean for fish to feed one. The federal government has given it all away to the foreign fleets."

The interdependence in the food chain was once acknowledged by the federal department responsible for Atlantic fisheries. In 1976, federal fisheries was a branch of the Department of the Environment (Fisheries and Marine Services Branch). Perhaps the association of bureaucrats responsible for both the environment and fisheries had something to do with the enlightened views about the interrelationship of creatures within the marine environment. The 1976 federal document titled "Policy for Canada's Commercial Fisheries" devotes an entire section to discussing the importance of "forage fish," or the feed stocks of commercial species:

"Forage Fish—In the past, stocks have essentially been managed in isolation from one another with little regard for the interaction between species. Total allowable catches (TACs) have been set on predators and prey alike without formally incorporating the interaction between species because of lack of knowledge of such interactions. There have been some exceptions to this; for example, the TACs for capelin have from the beginning been set below the estimated sustainable yield for capelin so as to allow for their importance in cod diet. In future, however, management of all species will need to consider these interactions. For example, if the strategy is to manage a predator species (eg. cod) to allow for its rebuilding, prey stocks (eg. capelin) will have to be managed to allow sufficient food for the ever-increasing predator biomass. This may involve under-exploiting the prey stocks to allow the surplus food for the predators. On the other hand, a prey fishery could be considerably more valuable than a predator fishery in which case some predator yield may have to be sacrificed in the long-term to fully exploit the prey species. These species interactions and their role in fisheries management will be the object of intensive research in the near future, research which has just commenced during the past couple of years."

Sadly, the appreciation expressed in 1976 was soon lost or subverted by those responsible for fisheries management in the newly formed Department of Fisheries and Oceans. The promise of "intensive research" into predator-prey relationships did not materialize, and the DFO proceeded with promotion of the exploitation, even overexploitation, of feed stock species. The Department has been willing to award enterprises with exploratory, developmental, and then permanent licences to fish feed stock species such as capelin, squid, shrimp, mackerel, herring, and silver hake without comprehending the impact such a fishery will have on commercial species. Hook and line fishermen have warned that such a foolhardy approach is destined to backfire.

The cyclical relationship between groundfish species and their food supply is not fully understood, even today. The interdependence is, therefore, often overlooked in the setting of catch quotas for target species—a contributing factor to the decline of groundfish stocks. Each species of fish tends to have its favourite food though some, like the cod, are not fussy beyond their first or second choices. Cod will eat most any food source available, whether it is bony fishes, free swimming shellfish, creatures which crawl along the seafloor, or attached bottom dwellers. It is this flexibility in diet which has made the codfish one of the most resilient groundfish species, able to adapt when its preferred choices for prey are in short supply. Due to its commercial popularity, more is known about the feeding habits of cod than other groundfish species, like haddock and pollock. Scientific study has traditionally hung on the apron strings of commercial development, and cod has been the undisputed industry leader of the Atlantic coast fishery since the days of John Cabot.

At first blush, it would appear that cod should not be wanting for a square meal. Yet, evidence shows that its favourite foods like capelin, herring, silver hake, and shrimp are also in trouble as they too are being targeted for commercial exploitation. The Newfoundland capelin fishery, for example, has been scaled back over the past few years after the closure of the offshore fishery in 1992. The herring fishery on the Scotian Shelf has been overfished and stocks are in rapid decline. Gulf of St. Lawrence herring stocks remain a mere shadow of their former abundance, although they are beginning to show some signs of stability. The inshore shrimp fishery has succeeded in wiping out all substocks thought to represent commercial potential, while modern factory freezers prepare to pursue northern and deep-water stocks.

Interestingly, quotas for these species including capelin, herring, squid and silver hake, have been scaled back by the federal government in a vain attempt to protect their commercial viability and not, as one might expect, to safeguard the integrity of vital food chains. Protection of forage stocks must be an essential part of the overall management, although, to date, those overseeing use of our ocean's resources have been very slow to acknowledge and embrace the concept.

Many of the feed stocks preferred by cod and other commercial groundfish have been termed "underutilized" by the federal government, meaning they have the potential to be more fully exploited commercially. As part of bilateral agreements with foreign fishing nations after extension of Canadian jurisdiction, these underutilized species were made available to the distant water fleets, which had an established history of fishing on Canada's east coast. Under these agreements, fleets were allowed allocations of stocks inside the 200-mile zone in return for a promise to cooperate with Canada in the management and conservation of resources, both inside and outside the zone. The acknowledged long-range goal for Canada was to claim an increasing percentage of these underutilized stocks, once considered surplus, and to "Canadianize" the entire fishery within the 200-mile zone.

The presence of foreign vessels in Canadian waters was never accepted by fishermen who argue that the entire resource should be preserved for the domestic fishery. The Department of Fisheries and Oceans attempted to downplay the significance of the foreign presence within Canada's 200-mile zone. But fishermen and Opposition politicians continued their attack on government policy. Why should huge factory freezer trawlers operated by foreign countries be allowed inside Canadian water to fish when the resources they remove could be used to ease the burden of rapidly declining domestic allocations of groundfish?

The most outspoken critic of government policy on foreign allocations was George Baker, Newfoundland MP for Gander-Grand Falls. During a debate on October 2, 1990, on a Conservative bill to increase fines for Canadian fishermen, the Liberal MP pointed out how easy it was for the fourteen Soviet and three Japanese factory freezer trawlers—fishing that day just off the Newfoundland coast and within the 200-mile limit—to receive permission to harvest capelin, redfish, and turbot stocks. "Fishermen are beginning to wise up to these licences that are granted by the Government of Canada. These foreign nations walk into an office in Halifax, Nova Scotia, or St. John's, Newfoundland, and over the counter they are given a licence for their vessel. They do not pay for it on the spot like Canadians do. They are billed later in the mail. They go out there in those massive vessels and destroy the fishing resource of Canada, using gear and machinery that is foreign to us...."

While the amount of fish allocated to other countries has decreased in recent years, the foreign presence in Canadian waters continues to grow. This trend is the result of the growing inefficiency of the Canadian offshore trawler fleet in the face of declining groundfish stocks and the profit margin involved in the pursuit of underutilized species. Domestic offshore companies have determined that it makes economic sense for them to farm out portions of their Enterprise Allocations for underutilized species to foreign fleets which have lower operating costs. Norwegian vessels, for example, were commissioned by Fishery Products International to catch turbot off Labrador in 1990. In the same year, the federal government permitted Bulgaria to move in for the first time and use its factory freezer trawlers to sweep the waters east of Cape Breton for mackerel.

The Government of Canada has implemented a directed fishery of such underutilized species as capelin, redfish, and turbot species, believing them to be surplus to the country's needs. Newfoundland longliner fishermen have fished turbot for decades, and capelin stocks are the focus of an important domestic fishery as well as a vital link in the food chain on which cod depends. Redfish has become an indispensable resource for fishermen and plant workers in the Gulf of St. Lawrence, where the decline in other groundfish stocks has meant the virtual shutdown of the industry. The list of stocks fished by foreign factory freezer trawlers goes on to include mackerel, squid, silver hake, plus over-the-side purchases of herring from Canadian vessels. Yet, these stocks are indispensable members of the food chain for commercially fished groundfish. Many fishermen maintain these fragile fish are not surplus. Pressure on these food sources by any fleet, domestic or foreign, has to impact negatively upon the health of commercial stocks already starved for a good meal.

While the directed fishery for these so-called underutilized species is cause for concern, the by-catch of commercial species like cod, haddock, and pollock poses graver consequences for already over-exploited species. The maximum approved by-catch of cod when fishing for turbot or silver hake is supposed to be 10 per cent. Should a foreign vessel find the by-catch level is higher, it is supposed to move to a different portion of the fishing grounds. Still, there is the by-catch that has initially been taken; the fish from that tow have died, having been crushed and squeezed in the huge net.

Suffering from the northern cod moratorium in 1992 and falling Canadian inshore allocations since then, fishermen expressed their frustrations with the continued foreign presence in Canadian waters. They blockaded a Russian cargo ship in port for one week in July 1993. A flotilla of about 100 inshore fixed gear vessels from southwestern Nova Scotia, who were tied up by reduced haddock by-catch limits, held captive the *Pioneer Murmana,* which was in Shelburne to offload frozen cod blocks scheduled for direct shipment to the United States. The protest drew national attention to foreign vessels fishing inside Canada's territorial waters. At first, fishermen maintained the fish had been caught in Canadian waters by trawlers and transhipped by the cargo vessel to port. Canadian and Russian authorities denied the allegation, saying the cod had been caught in the Barents Sea, north of Russia. These same fishermen had been made uneasy by the discovery, only one week earlier, of baby cod, hake, and pollock inside frozen blocks of squid bought from a Cuban vessel that had been fishing in Canadian waters.

Gary Dedrick, leader of the Shelburne blockade, addressing the Environment and the Economy Symposium on February 12, 1994, said fishermen could not understand how their own government could stop them from making a living and "...let foreign vessels inside our 200-mile zone, 50 to 60 mi. from shore, [catch] these very same fish ... that they were telling us we couldn't go for. The fish they were taking were ridiculously small. In fact, some areas of the Scotian Shelf were closed to Canadian fishermen from catching small fish, yet this did not apply to the foreign fleet...."

With the decline of the domestic groundfish fishery, there has been a push to encourage Canadian involvement in harvesting underutilized species. A shift to these fish food species is seen by government as an opportunity to take some strain off traditional groundfish species and at the same time provide fishermen and processing companies with an alternate source of income. This Atlantic Fisheries Adjustment Program was kicked off by Minister Tom Siddon with the 1990 Groundfish Management Plan, which included the establishment of a competitive pool of approximately 100,000 tonnes of underutilized species to be placed at the disposal of Canadian fishing companies and fishermen.

Two workshops, held late in 1989 and early in 1990 in Iqaluit, Northwest Territories and Yarmouth, Nova Scotia, were a follow-up to the establishment of the competitive pool. At these meetings, the significance of species such as silver hake, Atlantic mackerel, shrimp, and various shellfish, was promoted in the diversification potential of the east coast fishery. This direction was reinforced in May 1990, with new Fisheries and Oceans Minister Bernard Valcourt's announcement that $50 million in funding would be provided under the Atlantic Fisheries Adjustment Program to assist commercial ventures with start-up funding.

The Atlantic coast industry was not prepared for Valcourt's next move, which raised a storm of protest, the likes of which had not been heard since the restructuring of debt-ridden offshore companies in 1983. The pressure was on the federal and provincial governments to provide some relief following a series of plant closures in 1989 and 1990. Two communities, among the many reeling from these closures, were Canso, Nova Scotia, and Burgeo, Newfoundland. Both offshore-based towns were dependent upon National Sea Products, and with their plant closures in 1989, local politicians lobbied long and hard for a plan to save their communities. After weeks of rumours, a deal was announced on August 10, 1990, providing for the purchase of the Canso and Burgeo plant operations by Seafreez Foods Limited. As a company, Seafreez was a virtual unknown, but its principals—the Blades family of Nova Scotia and the Barry family of Newfoundland—had a long-established presence in the fishing industry.

Minister Bernard Valcourt indicated the deal was made possible under the framework of the Atlantic Fisheries Adjustment Program, with an allocation of underutilized fish and a loan guarantee of $6 million through a commercial bank. The deal meant 126,000 tonnes of underutilized species would be made available to Seafreez, including 50,000 tonnes of silver hake (whiting) on the Scotian Shelf and 50,000 tonnes of capelin off Labrador and eastern Newfoundland. The capelin award represented a whopping 40 per cent increase over Atlantic coast catches at the time. These two allocations were to be made from the total allowable catches set aside for Canadian interests but traditionally not harvested, while 26,000 tonnes were found in the new competitive pool. In addition, 19,990 tonnes of various groundfish species were transferred to Seafreez from Enterprise Allocations belonging to National Sea.

As previously mentioned, the deal was termed the "Canso bail-out" by the Nova Scotia industry and was a good deal for National Sea also, which was able to shed a $17.5 million debt owing on the Canso plant and pick up 2,000 tonnes of whole shrimp quota to be processed at its plant in St. John's. National Sea still knew its way to the trough, despite the lapse of nearly ten years since its last major bail-out.

The deal was equally beneficial for Seafreez. Along with the allocation of underutilized species, it picked up the Enterprise Allocations from National Sea, a loan guarantee from the federal government, and an agreement from the Province of Nova Scotia to carry the $17.5 million debt for the next three years. After that, the province would write down the debt by $1 million a year until it was paid in full as long as Seafreez employed 500 people for ten months of the year at the Canso plant. In addition, the Atlantic Canada Opportunities Agency (ACOA) granted Seafreez as much as $5.5 million to upgrade refrigeration and freezer facilities at the Canso plant.

The deal also allowed Seafreez to trade its allocation of underutilized species for groundfish caught by the Russians outside Canada's 200-mile limit. The domestic industry was having difficulties making the harvest of capelin and silver hake a viable venture. The association with the Russians in the commercial exploitation of these species was to be a learning experience to help pave the way for eventual Canadianization of the fishery. The trade was criticized for granting too much control over the harvest of underutilized species to the Russians, who were already the most dominant force in the silver hake fishery. The competitive pool of 100,000 tonnes of underutilized species had been created for the benefit of Canadian interests and to encourage Canadianization of the fishery, but a large chunk of the available pool had been captured by the Russians. The result would be that more of the groundfish feed stocks would be taken for human consumption or for the production of fish meal, jeopardizing the food supply for groundfish stocks, and ultimately stock health and replenishment.

The favoured treatment for Seafreez soon came to an end as the company showed itself unwilling or unable to meet its obligations stipulated under the agreement. Under the Atlantic Fisheries Adjustment Program, Seafreez had committed itself to continue operation of the two plants, but in 1992 Seafreez closed the Burgeo plant. Seafreez had also committed itself to the processing of specific percentages of fish at each of the Burgeo and Canso plants. The silver hake and capelin allocations were to be traded for Barents Sea cod and shrimp harvested by Russian vessels outside the 200-mile limit. The Enterprise Allocations were to be processed equally at the two plants, while the underutilized species and bartered fish were to be shared on a 60/40 basis by Canso and Burgeo. Subsequent investigation by federal authorities showed that none of the underutilized and bartered fish were processed at Canso. In addition, 1,250 tonnes of bartered fish caught by the Russians in Canadian waters were sold directly to third parties. Furthermore, all of the groundfish by-catch of the Russian fishing operations was to be processed at Canadian plants, but Seafreez landed less than half of it.

Three years after signing the deal, Minister John Crosbie terminated the special arrangements, citing several breaches of contract. On March 19, 1993, Minister John Crosbie voided Seafreez's barter allocation of silver hake and capelin. He also withdrew Seafreez's special status for underutilized species allocations, requiring the company to apply on an equal footing with the rest of the industry.

The Seafreez deal, more than any other program announced by the federal government, showed a lack of appreciation for the delicate balance of the marine food chain. Capelin were, and still are, classified as underutilized. The federal government had not looked to the Barents Sea fishery and had not listened to the expert advice it had commissioned in its study of the decline of northern cod stocks. Capelin stocks in the Barents Sea had been fished nearly to extinction, with a 1 million tonne quota in the mid-1980s reduced to zero by 1990. With the collapse of the capelin, dramatic decreases in cod growth occurred, and some year classes were studied for relation to scarcity of its favourite food source. Cod quotas in the Barents Sea also dropped off the charts, falling from 600,000 tonnes to a mere 30,000 tonnes over the same ten-year period. This surely was not a coincidence.

When Leslie Harris appeared before the House of Commons Standing Committee on Forestry and Fishery on April 10, 1990, he explained what had happened to cause scientists and managers watching the Barents Sea to underestimate the impact of the fishing effort. "Their [cod] weights at age were wrong for another reason that we have not seen in our case yet. Their weights at age were wrong because they overfished their capelin population enormously. They began using capelin as the feedstock for a meal industry, and they fished the capelin horrendously hard. They depleted them to the extent that there were none available to feed the cod, because capelin make up the principal prey species for cod. Without food, the cod began to lose weight and to fail to grow at the same rate as they should have grown at."

In light of the Harris Report released in February 1990, it seemed the current Minister of Fisheries and Oceans, Bernard Valcourt, was courting disaster with the Seafreez deal. The 50,000 tonnes of capelin originally given to Seafreez and representing a 40 per cent increase over Atlantic catches was sure to have a great impact on the health of cod and other predator species.

Although the Seafreez special deal for capelin and silver hake has been trashed, the attitude of the federal government remains about the same. Despite the change to a Liberal administration, federal authorities are still eager to give away feed stocks no matter how crucial to the food chain, as long as they will create or support short-term jobs. Witness the continued plunder of shrimp stocks off Newfoundland and Labrador and the promotion of the silver hake fishery in Nova Scotia. Often the jobs are in plants owned by the large offshore companies, the same companies that have been well positioned to appeal to the fancy of fisheries managers and politicians.

A respite was given for capelin off Newfoundland when Fisheries Minister John Crosbie announced, months prior to the moratorium on northern cod fishing, that the 1992 offshore capelin fishery would be closed. A review of the state of the capelin stocks showed they were in serious trouble. Though it has been scaled back, the inshore capelin fishery for Newfoundland continues. The 1993 total allowable catch was set at 45,390 tonnes, a decrease of 12,945 tonnes from the previous year. The impetus for this reduction was not to preserve the vital link in the food chain but to protect future commercial endeavours directing for capelin.

Fishermen plying the waters off Cape Breton suspect the Department does not recognize the connection between capelin and cod stocks. A March 1994 letter from Local 6 of the Maritime Fishermen Union indicates their apprehension. Secretary-Treasurer Jeff Brownstein indicates that inshore fishermen have good reason to believe that Fisheries and Oceans Minister Brian Tobin has not made the connection between cod and capelin. Referring to a rumour that a company had recently received a 1,000 tonne quota of capelin, his remarks appear in the *Sou'Wester* of March 15, 1994: "To add to the absurdity of all this, we are wondering how anyone could propose a capelin fishery when the cod fishery is completely shut down. Everyone knows that capelin are the most important food source for so many species of fish—particularly cod. In countries like Norway where the cod stocks collapsed, they stopped fishing for capelin and the cod stocks rebounded. Whatever happened to erring on the side of conservation?"

Silver hake, or whiting as it is often called, is also left open to attack by the Russian and Cuban fleets which have been hammering away at the stocks off Nova Scotia since 1977, when they first reached a bilateral agreement with Canada. With their low operational costs, the foreign fleets have been able to profitably take the slender fish using small trawler meshes and produce a salted, canned, or frozen product for consumption in their home countries.

Silver hake was touted as the most abundant groundfish on the Scotian Shelf with an estimated biomass of 1 million tonnes. Interest from the Canadian industry surfaced in 1988, and by the next year, the first domestic catches were being recorded. In March 1990, Valcourt announced approval for five Canadian proposals to harvest and process a total of 35,000 tonnes of silver hake on the Scotian Shelf. The new projects were to provide at least 225 direct jobs with about 200 of these designated for the closed Lockeport National Sea plant. The stated intention of the silver hake program, on the part of the federal government, was to divert some of the pressure exerted by the inshore dragger fleet away from traditional groundfish stocks. Also, the job creation for the Lockeport plant fell through when it was flipped to Clearwater Fine Foods by the Nova Scotia Conservative Government. This back-room deal was soon forgotten in the furore created by the 1990 Seafreez deal for the Canso and Burgeo plants.

When development of the silver hake sector stalled, the federal government opened up the purse strings, showing its desperation to get this new segment off and running. In 1992 Blue Wave Seafoods of Port Mouton, Nova Scotia, was given a grant of $240,000 and a repayable loan of $245,000 to establish an operation dedicated to production of value-added products, ranging from butterfly fillets to mince blocks. The inshore dragger fleet was expected to provide the raw product but the enterprise had difficulty getting off the ground.

Seventeen Canadian proposals for silver hake harvest were accepted for 1993, calling for 73,000 tonnes to be harvested by enterprises throughout the Atlantic region. This total eclipsed foreign allocations set at 13,000 tonnes. A further reduction in the foreign take was announced in April 1994, when the total of all underutilized species, including silver hake, was reduced to 9,700 tonnes from the 30,700 tonnes allowed in 1992.

While foreign involvement in the silver hake fishery was being phased out, efforts to turn it over to the domestic sector were stepped up. Early in 1994 the Committee for the Canadianization of the Silver Hake Fishery was established with the expressed goal of developing a home-grown industry. Again Blue Wave Seafoods of Port Mouton, also represented on the committee, was to be the chief beneficiary of the lobbying of the federal government. Representation on the committee also included spokesmen for inshore dragger fishermen, DFO senior advisor Dave Lemon, and South Shore, Nova Scotia, Liberal MP Derek Wells. The committee boasted that by-catch in the silver hake fishery had been reduced to negligible levels. The committee failed to acknowledge that the by-catch species not taken by the net were often mortally wounded by the separation grate used in the process. The committee also shied away from any suggestion that a commercial harvest of silver hake would further delay recovery of the higher valued groundfish stocks which depend on silver hake for food.

The burgeoning shrimp fishery is parallel to the silver hake fishery. Shrimp tend to form a significant portion of the diet of cod from an early age, though their importance tends to vary in direct proportion to their availability. When shrimp is in short supply, cod will shift to another food source. Shrimp are also a favoured human dish, and fishermen, with the encouragement of the federal fisheries department, have been quick to pounce on any commercially attractive populations as they appear.

The popular commercial species, *Pandolus borealis,* found throughout the region from the Gulf of Maine to Labrador, has attracted particular interest. These stocks have been pursued for the past thirty years by inshore fishermen who, recognizing the opportunity for a fast buck, quickly converted their vessels to fish the species. As each stock was fished out, the boats retrofitted to fish the crustaceans were again geared up to drag for groundfish. Stocks all along the Atlantic coast have fallen victim to a pulse fishing effort whereby the mobile fleet moves from concentration to concentration of the shrimp, fishing it down and thus preventing the stocks from growing to their stable potential.

In most instances, only a technicality—that is, the associated by-catch of undersized groundfish—prevented the shrimp from being completely fished out. This was not a small deterrent, however, as shrimp vessels around the world have taken countless immature groundfish and have thus inflicted an immeasurable toll on stocks. The meshes used to catch shrimp were only 40 mm and also trappped the small immature groundfish feeding on the food stock. The meshes have caught very small specimens of about 28 g (1 oz.). The by-catch of groundfish needed to cease as the troubling question persisted: Why were we fishing the food sources of our higher valued groundfish?

The development branch of the Department of Fisheries and Oceans was instructed to work on a solution to the by-catch problem associated with the shrimp fishery. With the lead work completed by Scandinavian scientists and technicians, the development branch was able to adopt the Nordmore grate, introduced in the late 1980s, for use in the Canadian shrimp fishery. The device traps the shrimp, while permitting an escape route for larger fish. Trials have shown that the metal grate can reduce the by-catch of groundfish to less than 5 per cent, thus meeting allowed levels. With this new gadget, DFO felt justified in promoting the shrimp fishery. However, the Nordmore grate was not mandatory for Canadian vessels until 1993, and it is still subject to further refinement.

The glamour in the shrimp fishery lies with the large offshore vessels fishing the northern waters off Newfoundland, Labrador, and Greenland. These are the same grounds which serve as a nursery area for northern cod and thus are part of the base in the food chain. The companies involved in the shrimp fishery are the same ones in our offshore sector which have inflicted the greatest direct impact upon our groundfish stocks. Nearly twenty Canadian offshore shrimp trawlers fish the stock inside the 200-mile zone, while outside the foreign fleets are ready to strike should any commercial concentrations be identified outside Canadian jurisdiction.

During the summer of 1993, word travelled the world that a huge find of shrimp had been made on the Flemish Cap, beyond the Grand Banks off Newfoundland and just outside Canadian jurisdiction. This is the same area that scientists involved in surveying northern cod populations believe is the final retreat of the stock. Excellent catches of shrimp as high as 159 tonnes a week were being reported, as an international fleet of more than thirty world-class trawlers converged on the spot, most of which had sailed from the Norwegian prawn grounds. *Fishing News International* of August 1993, reported there were Norwegian and Canadian vessels as well as ten ships from Faeroe Island and five from Iceland.

The shrimp were soon depleted—earlier catches of 20 tonnes per day were reduced to between 6 and 8 tonnes. Even the smaller catches would fill the deep freezing holds of most of the Icelandic vessels in twenty days of fishing. Plans called for a huge freighter to deliver packaging material and receive the shrimp at sea, so the trawlers could continue fishing the grounds without having to return to their home port for supplies. This tag-team approach to fishing, now practised by nearly all the world's major fleets, indicates the power of the infrastructure ready to exploit any stock.

The Canadian government responded to this raid by foreign vessels in the typically Canadian way, politely asking foreign countries to keep their shrimp fishing boats off certain portions of the fishing grounds to avoid the by-catch of northern cod. The government received no official reply from members of the Northwest Atlantic Fisheries Organization, and the issue fizzled out, further exposing Canada's impotence when dealing with marauders on the high seas.

Mackerel, known to be a choice feed stock of groundfish, is also being promoted by the Canadian government for exploitation. Inshore trap net and gillnet fishermen have traditionally taken mackerel to provide bait for the hook and line fishery. Even with these passive methods, mackerel catches have shown continued decline, indicating poor stock health. The new initiative by the federal government has been to encourage the highly efficient purse seiner and dragger fleets to enter the mackerel fishery. In 1992, Fisheries Minister John Crosbie announced three-year allocations to nineteen Canadian companies to further develop the mackerel fisheries. "These projects will provide Atlantic Canadians with valuable harvesting and onshore processing employment," he explained in a DFO press release of January 11, 1993. "By developing this fishery, we hope to acquire both the knowledge and expertise to diversify the Atlantic fishing industry." The stated goal was noble, but the method to be used was full of pitfalls.

Robert Conrad, President of the Nova Scotia Mackerel Association, warned in *The Atlantic Fisherman* of March 1993 that an enlightened approach was desperately needed. "Respecting the mackerel stock itself, the education of fishermen, fishery policy makers and politicians must continue. Resources are limited and must not be wasted.... We are woefully short in our comprehension of how the mackerel biomass interacts with its environment." He cautioned, "The same machinery that's kept the dragger fleet scouring our ocean bottom is about to attach its tentacles to the mackerel biomass. We need to fear, and we need to act." It was earlier that year (January, 1993) that Minister John Crosbie had awarded $37,500 in funding so that Blue Ribbon Seafoods, a subsidiary of Clearwater Fine Foods, could construct a facility to process underutilized fish species, including mackerel.

The squid fishery is yet another example of the exploitation of groundfish food sources. Fishermen still recall when the species was found in abundance on the squid jigging grounds. An international fishery for squid off the Atlantic coast developed in the late 1960s and early 1970s. In 1976, catches climbed dramatically and remained high until 1981, when a stock crash occurred. Possibly coincidence, but the decline and continued failure of the squid stocks has been parallel to the failure of northern cod stocks and other groundfish along the Atlantic coast. Squid stocks on the Scotian Shelf rebounded somewhat in 1990. This resurgence was short-lived, however, as they were quickly seized upon by Japanese jiggers licensed to go after the resource. Soviet and Cuban fleets also took squid while directing for silver hake in waters south of Nova Scotia.

Herring is one crucial Atlantic coast feed stock that no one considers underutilized, having served for thirty years as the basis of a flourishing industry. While the herring fishery continues to attract heavy investment today, its legacy of rusting onshore infrastructure indicates that its power has often been misdirected. Herring concentrations have been unable to withstand the aggressive effort of the highly mobile purse seining fleet, and subgroups of this valued pelagic species have been reduced to remnants of what they were. The stocks, which were considered healthy in the late 1960s, were subjected to the boom in fishing power through the 1970s.

The Gulf of St. Lawrence herring stock was one of the first to go down. In 1965, purse seiners were brought in from the Pacific coast, where that resource had been depleted. The fleet began fishing in the southern Gulf and along the southwest coast of Newfoundland where the adult herring overwintered. The take increased to 300,000 tonnes in 1970. The fishery could not withstand the pressure and "suddenly" collapsed in 1974 with landings at 37,000 tonnes.

This story was repeated in the Bay of Fundy, where the traditional gillnet and weir fisheries were overrun by purse seiners in the 1960s. The growth was driven by the developing fish meal business. The purse seine fleet expanded from 20 to 107 vessels between 1963 and 1968, and total purse seine landings jumped from 22,000 tonnes to 225,000 tonnes. This stock also crashed with 65,000 tonnes taken in 1971. Today, communities such as East Pubnico and Yarmouth bear the scars with abandoned factories presenting an eye-sore for passers-by.

Other herring fisheries have had similar fates. As well as the Bay of Fundy and Gulf of St. Lawrence fisheries, those on the east coast of Newfoundland, Placentia Bay, Sidney Bight, Chedabucto Bay, and Georges Bank, have also declined. The latter fishery was very productive in the 1960s, with overfishing in 1968 netting 373,344 tonnes. No commercial herring fishery exists today, and there are only preliminary signs of stock return.

On the international scene, the North Sea herring fishery is perhaps the most noted failure which also followed the introduction of new technologies. Purse seiners, driven by more powerful engines, sophisticated hydraulics, and fishfinding electronics, forced the collapse of this fishery in the early 1970s. This took groundfish stocks down as well, exposing the dependence of cod and related species on a stable food supply. Because herring, like other pelagic species, have a natural tendency to school, they are especially vulnerable to mobile fishing effort. The British Columbia and North Sea experiences showed that herring, under increased fishing pressure, form smaller and smaller schools as they crowd into reduced regions of their known range. With continually improving fishfinding equipment, purse seining fleets are able to locate these concentrations and maintain a respectable catch rate until the stock is forced into irreversible decline.

While herring stocks throughout the Atlantic region remained at depressed levels, development of the herring roe market, with strong prices through the mid-1980s, meant renewed interest and renewed pressure. In the Northumberland Strait, a herring roe fishery was developed with a 10,000 tonne quota. Renewed fishing pressure forced a collapse in 1991, with stocks down and prices also on the floor. The season experienced a similar low in 1992.

Stocks in the Scotia-Fundy region, which includes the Bay of Fundy and waters to Cape Breton Island, showed some signs of stability despite the establishment of a herring roe fishery through the 1980s. By the end of the decade the telltale signs were evident. Under-reporting and failure to report catches by the seiner fleet were blamed for the disastrous results and breakdown of a ten-year plan to rebuild herring stocks to former levels in the 1960s. In 1989, DFO introduced new regulations for accurate reporting of herring catches and moved to expand the already closed spawning area known as Trinity Ledge off Yarmouth County. The closed area was increased from 91 km^2 (35 sq. mi.) to 259 km^2 (100 sq. mi.) because authorities were concerned that few spawning herring had returned.

The new regulations went largely unheeded and while fishermen off southwestern Nova Scotia reported a catch of 77,500 tonnes in 1990, the amount of herring and herring products reaching market indicated the catch was more in the order of 120,000 tonnes. DFO asked fishermen and companies to restrain themselves. While acknowledging biological advice to reduce the quota, the department set the 1991 quota at the previous year's mark, hoping that the industry would see the necessity for responsible reporting and responsible fishing. The effective correction was not administered by DFO but imposed by nature. The roe season in 1994 was largely a bust due to the failure of full-sized, mature spawners to return to the grounds. Overfishing had taken its toll again.

Canada is not the only nation responsible for encouraging overdevelopment of feed stock fisheries. Although the once great anchoveta fishery of the southeast Pacific was reduced from a yield of 10 million tonnes in 1970 to nearly zero in 1973—the most spectacular collapse in the history of fisheries exploitation—the Vice-Minister of Fisheries for Peru recently invited world fishing interests to help his country upgrade the fishing effort off Peru's coast, to take advantage of "underdeveloped" species, such as horse mackerel, chub mackerel, and squid. He issued the invitation, at the same time acknowledging that the sardine and remaining anchoveta stocks were still under pressure from overfishing and had not recovered from the crash of the 1970s.

It is apparent that governments would rather ignore the concerns of fishermen and industry representatives, when their caution casts doubt on an agenda for development. The federal government was given ample opportunity to hear and respond to the fears of fishermen who warned of the imminent collapse of groundfish stocks. All the warning signs were there for fishery scientists and managers to heed. Fish numbers had fallen and were continuing to plunge; average fish size had shrunken dramatically, with the proportion of older, mature, and more fecund fish continuing to decline; and groundfish were exhibiting low weight-at-age traits which indicated they were not getting enough to eat. The industry brought these concerns to government. They warned of the consequences of a build-up of fishing capacity; of overfishing; of the use of trawls and the resulting destruction of marine habitat; of destructive practices of highgrading, discarding, and dumping; and they warned of the danger of fishing feed stocks on which groundfish are dependent. These troubling signs were presented to government in formal and informal settings. The concerns of fishermen and industry representatives were made known to the Hache Task Force and the Harris Task Force. At every turn, apprehensions about the future of the Atlantic groundfishery were overshadowed by interests more concerned with short-term industry stability and corporate profit than with long-term stock sustainability. The "drift with the tide" approach taken by government ensured the inevitable result—the crash of groundfish stocks and the closure of the fishery.

Chapter Nine

Fish or Fishermen First: Tobin's Ecosystem Approach to Fisheries

The enormity of the crisis in the Atlantic groundfishery and the responsibility for recovery weighed heavily on the new Liberal administration, elected in the fall of 1993. According to *The Chronicle-Herald* of July 25, 1994, after only six months in his new office, Fisheries and Oceans Minister Brian Tobin was ready to announce a fundamentally new approach to fishery management. He indicated the groundfish crisis had prompted his department to re-evaluate management objectives; the self-evaluation had transformed DFO, and not a moment too soon. Tobin said both industry and the government must share blame for the collapse of groundfish stocks because previously, "there was an ethic in the fishing industry that said fish should conform to people's personal income ambitions, people's business plans, provincial aspirations." The system was being driven in reverse with industry priorities set first and fish allocated second, instead of determining how much fish was available and making business plans to match the resources.

According to the new approach of DFO, the fish would come first. If it was obvious that stocks were declining, Tobin promised he would not hesitate to protect the stocks. "And if that means some people's expectations are dashed, if that means abrupt change, if we look less than smooth, than I guess that's the way we're going to have to look." Stock health would finally receive deserved attention, shifting the focus away from short-term maintenance of the fishery. The harsh reality for fishermen was the change to a "fish first" policy would mean that many would be forced from the fishery. With their boats tied up indefinitely or their operations greatly scaled back, many have been forced to sell out their operations or face bankruptcy. The cost of belated government enlightenment for east coast fishermen is unemployment. Eleventh-hour measures—more akin to calling in the undertaker than calling in the physician—are cold comfort to those watching the demise of their way of life.

The new management policy of fish first—an ecosystem approach—had been the main emphasis in a bevy of recommendations issued by the Fisheries Resource Conservation Council (FRCC), established in January 1993 by the previous Conservative administration. The FRCC Report of January 28, 1994, states: "Thinking in terms of the whole ecosystem must become an essential and integral part of day-to-day activities, not just for Science, but within the Department of Fisheries and Oceans generally." More emphasis should be placed on the study of the marine ecosystem and less on the establishment of quotas. Displaced fishermen were told that they should welcome the new direction. Though there would be tough times ahead, government was doing the proper thing by focusing on protection of the remaining groundfish and taking measures to encourage stock recovery. There would be an adjustment program for fishermen and plant workers to ease the strain. The industry was assured the tough cuts would be worth it.

Tobin had already indicated his willingness to embrace the new philosophy promoted by the Fisheries Resource Conservation Council when he moved to close the groundfishery throughout most of Atlantic Canada in December 1993. Pointing to the need for conservation and caution he closed all but one cod fishery—that off southwestern Nova Scotia—and severely reduced catches of other species, putting another 5,000 people out of work. Tobin was following, and in some cases exceeding, advice given to him by the FRCC. He defended his decision to go beyond some of the council's recommendations, stating in *The Daily News* of December 21, 1993, that he would rather err on the side of conservation than allow fish stocks to plummet further.

Tobin has followed through on his commitment to favour the fish, even in the face of industry opposition. He has shown that he is willing to scale back or even close a fishery. Some of these fisheries may be closed into the next century if that is what it takes to ensure a return to stock health. With key commercial species showing little resilience, prolonged closures appear inevitable.

While closures and quota cuts have hurt the domestic fishery and generated criticism for Tobin, his move to confront Spanish trawlers overfishing turbot on the Nose and Tail of the Grand Banks and just beyond the 200-mile limit brought unequivocal praise from Canadian fishermen and much of the international community. In March 1995, a tough-talking Minister Tobin made it clear that he would not allow turbot to go the way of other groundfish, and he would not permit the European Community, Spain and Portugal in particular, to devastate turbot the way its members had ravaged other groundfish stocks.

On March 9, 1995, armed Canadian officials boarded and seized the Spanish trawler *Estai* and escorted it into port in St. John's, Newfoundland. The trawler's 5.5 tonne net was later recovered at sea and presented to the world via a well-orchestrated news conference held on a barge floating in

New York's East River. Tobin was in New York to address the United Nations Assembly on foreign overfishing and the need for coastal states, like Canada, to be able to protect straddling stocks, like turbot, which swim the waters inside and outside territorial lines.

Tobin's "turbot tussle" served to reinforce the Minister's commitment to a fish first policy and also acted as a temporary diversion from the very serious problems facing the domestic fishery. For more and more fishermen bankruptcy looms. Many are regretting their heavy investments near the end of the boom years, when they upgraded to newer and larger boats, often borrowing large sums and posting homes and vehicles as collateral. Others, involved in the dragger fishery, bought up boat quotas from retiring fishermen hoping to gather enough fish to make a living. With steady declines in cod stocks, Ottawa has cut back the amount of fish available under the quota system, making an increasing number of these operations marginal at best and often not financially feasible. Inshore fixed gear fishermen have also found it more and more difficult to land an amount adequate to meet expenses. Satisfaction and reward have been taken out of the fishery; most fishermen find it increasingly difficult to muster the courage necessary to set out for another day's fishing.

While the government of Nova Scotia has been lenient, allowing fishermen to fall behind in their boat payments to the Fisheries Loan Board, the same is not true of their Newfoundland counterparts. The Newfoundland Fisheries Loan Board has become edgy, often moving in on fishermen delinquent with their boat payments. An increasing number of judgements have been granted by the Newfoundland Supreme Court, permitting the authorities to take action to secure personal property posted as collateral. Unable to honour their obligations, fishermen are having to stand aside while their family investments are placed on the auction block. Ellis Crocker of Green Harbour, Newfoundland, owed approximately $60,000 and expressed fear that he would lose his home if the judgement was executed. "They want it all right now and I'm getting one cheque in here now which is $600 every two weeks," he told *The Chronicle-Herald* on February 2, 1995. Crocker received support from The Atlantic Groundfish Strategy (TAGS) program set up to assist displaced fishery workers with income support and retraining programs.

While the northern cod fishery off Newfoundland has been shut down since the summer of 1992 and much of the remainder of the Atlantic fishery since 1993, there remains a considerable proportion of industry workers who are either unwilling to accept the realities of the crisis or are simply reluctant to respond. According to *The Chronicle-Herald* of June 21, 1993, Tom Best— a fisherman from Petty Harbour, Newfoundland, and President of the Association of Newfoundland and Labrador Fisheries Co-ops—says "The vast majority think the fishery will come back.... So it's very difficult to get them to take advantage of government paid training opportunities right now.

We consider this very dangerous." The response along the coast is similar with fishermen unwilling to enrol in training options made available in most localities. Many assume if they enrol in training options, they may jeopardize their chances to remain in the fishery. They fear acceptance of a retraining option will be used by government as an excuse to force them from the fishery.

Boredom has replaced work for many fishermen located in the 700 communities of Newfoundland and for those in an increasing number of shutdown ports throughout the Maritimes. There is only so much time that can be spent repairing and storing fishing gear. The rest is spent pacing the floor or reminiscing about the good old days when it was a pleasure to get up before dawn and set out for yet another fishing trip. Similarly, for many former fish plant workers, adjustment comes slowly. Some remain hopeful that their plant may one day re-open, though the prospects become more bleak each day.

A more perilous situation sometimes results when unemployed fishermen and fish plant workers find themselves spending increasing amounts of time at home. Stress, brought on by drastic cuts to the region's fishery and the loss of income, has given rise to a dramatic increase in family violence in many coastal communities throughout Atlantic Canada. A Newfoundland report, completed for the Fishermen, Food and Allied Workers Union, involved a three-month tour of twenty-seven communities in that province. Barbara Parsons, author of that report, said she found many women experienced beatings by stressed mates who had turned abusive because of the crisis in the groundfishery. "There's a lot of frustration out there," Parsons told *The Chronicle-Herald* on November 8, 1993. "In these small communities, there are people lying around, people with time on their hands, alcoholism, violence—the whole shaggin' bit."

A similar situation is reported in *The Chronicle-Herald* on April 23, 1990. In southwestern Nova Scotia, the administrator of a women's shelter in Yarmouth explained that unemployment and mounting financial worries aggravate abusive relationships. Pam Decker of Lockeport, Nova Scotia, Chair of the Shelburne County Women's Fish Net, says her group is dedicated to "helping our husbands fight for their survival in the fishery ... and for survival of our families and our communities." But Decker is concerned that recent statistics released by Shelburne RCMP indicate assaults are up significantly in that jurisdiction, and spousal assaults are up from three recorded in 1990 to fourteen recorded in 1994. Shelburne County Fish Net is part of an information and action network of women's organizations being developed across Nova Scotia to enable women of fishing communities to share their concerns and experiences.

Individuals have been encumbered by the weight of the crisis and the economies of coastal communities have, in many cases, been decimated. Most often these communities were dependent upon the fishery and often isolated from other sources of industry and commerce. This is especially true for the many fishing communities spread out along the ragged coast of Newfoundland. Public leaders there compare the present day pressures to those that led the Joey Smallwood government of the late 1960s to institute a resettlement program to larger population centres. Hardship has affected municipal treasuries, and property owners are left with little choice but to renege on payment of their taxes as welfare rolls continue to swell.

After first denying that there was a crisis in the fishery and then procrastinating for more than a year, the federal Conservatives unveiled their major aid package for the Atlantic coast on May 7, 1990. The $584 million federal initiative, called the Atlantic Fisheries Adjustment Program, provided funding for rebuilding fish stocks ($150 million); adjustments to "current realities," or a retirement package for older fish plant workers and trawlermen affected by plant closures ($130 million); and economic diversification funding ($150 million). Of the latter sum, $50 million were dedicated to promoting economic diversification within the fishery and were to focus on diversification in the areas of fisheries development, maximizing use of underutilized fisheries, marketing, and aquaculture.

In November 1990, the federal government announced a program for diversification outside the fishery. The Fisheries Alternatives Program (FAP) had funding of $90 million and was administered by the Atlantic Canada Opportunities Agency (ACOA) on behalf of the federal government. The five-year FAP was designed to create permanent employment for displaced fishery workers and to diversify the economies of coastal communities. In fact, the program was a shameless attempt by Ottawa to attract interest away from the groundfish industry. Money was made available for the establishment of new businesses, research and development activities, the expansion or modernization of enterprises, and the development of business infrastructure. Activities eligible under FAP included manufacturing and processing operation (except the fishery); tourism projects, including accommodation and attractions; industries providing service to business, and businesses that serve the primary industries; commercial research and development; and other sectors considered capable of contributing to the economic diversification of areas affected by the downturn in the fishery. Almost every enterprise was eligible for funding.

Much of the money was squandered, handed out indiscriminately by federal bureaucrats driven to meet dispersement quotas. Politicians, under attack for dragging their feet and showing little initiative, cut their underlings loose. They were anxious to prove that they were doing all within their power to run the industry and keep local economies afloat. The haste made for poor judgement.

A federal audit of seventeen major projects administered by ACOA, the agency responsible for administering the Atlantic Fisheries Adjustment Program, showed the body does not have a particularly good track record. The audit's findings appeared in *The Chronicle-Herald* of September 6, 1995, and showed that the federal development agency "routinely failed to spot trouble when entrepreneurs made pitches for federal grants, ... marketing plans weren't challenged, cautions from experts were ignored, and projects were never monitored to ensure taxpayer's money was well-spent." The audit zeroed in on seventeen projects administered by ACOA between 1987 (before FAP) to 1992, all of which failed and left taxpayers on the hook for $100 million. Money was thrown at ill-advised projects and the goal of long-term diversification of coastal communities failed to materialize. As a result the downward spiral intensified.

The plunge hit bottom with the crash of northern cod stocks and the imposition of a fishing moratorium in July 1992. The federal government was quick to react, announcing at the July 2 press conference the Northern Cod Adjustment and Recovery Program (NCARP), designed to assist about 20,000 fishery workers affected by the closure. At first, fishermen had to content themselves with an income supplement of $225 a week in emergency funding until a more comprehensive plan could be devised. By March 1993, 22,000 declarations of intent had been received from Newfoundland fishermen and plant workers seeking qualification for $406 per week. To qualify for the maximum benefits, recipients had to choose one from a number of programs offered, including early retirement or skills training. Nearly 13,000 workers opted for training within the fishery signifying their desire to stick with a familiar vocation, one which had served them well over the years. Only 2,500 indicated a willingness to train for jobs outside of the fishery. "Retrain for what?" was a common line. An additional 1,900 workers chose early retirement, realizing the moratorium could extend into the next century.

A federal audit of the Fisheries Alternatives Program and its successors, conducted in October, 1994, showed that few Newfoundland fishery workers benefited from the $115 million fund set up for that province. Only 257 of the 16,000 displaced fishery workers actually found gainful employment, while the remainder of the 3,400 jobs created in Newfoundland went to the unemployed from other sectors of the economy. The report indicated that fishery workers were not prepared for career change as a result of an often early exit from the public school system. *The Chronicle-Herald* reported on January 26, 1995, that only about one-third of fishing industry participants had achieved a high school diploma compared to two-thirds of those employed in the general workforce.

Similar responses to the training options came from fishermen and plant workers with the extension of groundfish closures to the Gulf of St. Lawrence and eastern Nova Scotia in August 1993. With the added closures, the federal government expanded transitional assistance to include 9,000 to 12,000 fishermen and plant workers, in addition to the 25,000 people left unemployed by the northern cod moratorium. The expanded coverage, known as the Atlantic Groundfish Adjustment Program (AGAP), cost the federal government an additional $190 million in assistance, which was provided not only for retraining but also to help fishermen cover loan payments for their vessels. The transitional assistance for Gulf and eastern Nova Scotia fishermen was modelled on the Northern Cod Adjustment and Recovery Program, and both plans were set to expire on May 15, 1994.

On April 19, 1994, Tobin announced a comprehensive five-year program of adjustment and income support to assist an estimated 30,000 fishermen and fish plant workers across the Atlantic region and to replace the two existing programs—Northern Cod Adjustment and Recovery Program and Atlantic Groundfish Adjustment Program—which were due to expire. The program cost the federal treasury $1.9 billion and was titled The Atlantic Groundfish Strategy (TAGS). TAGS was intended for industry adjustment, designed to reduce capacity and downsize the fishery. The federal strategy, boldly stated, was to achieve capacity reduction in all sectors of the groundfishery with a goal of 50 per cent reduction. In making the announcement, Tobin cautioned that governments would have to show resolve and the fishing industry would have to be ready to address the issues of harvesting and processing overcapacity in the Atlantic groundfishery. Tobin's remarks appeared in the *Sou'Wester* on May 1, 1994: "An open dialogue and partnerships among governments, industry, and fishermen's organizations in the coming weeks will ensure that structures are established to bring about industry renewal in a manner that fully reflects local concerns and conditions.... This will allow us to begin the process of rebuilding an economically viable and environmentally sustainable fishery of the future." To accomplish this, the "Industry Renewal and Capacity Reduction" component of TAGS would involve the establishment of industry renewal boards. These quasi-governmental bodies were first floated during hearings conducted by the Task Force on Incomes and Adjustments in the Atlantic Fishery and adopted as a cornerstone recommendation in the resulting Cashin Report of November 1993. With industry involvement on these Industry Renewal Boards, the federal government was looking for sanction of its capacity reduction plan.

While many fishermen and plant workers welcomed TAGS assistance, others were leery of its implications and concerned for the federal government's apparent agenda. Would acceptance of assistance under TAGS come with strings attached? Would industry participants be required to retrain for work outside the fishery? The federal government had repeatedly indicated there were too many fishermen chasing too few fish. This was the same line from former fisheries ministers like Bernard Valcourt who had, in turn, dusted off the theme of the Kirby Report of 1982. Michael Kirby pointed out the overcapacity in the groundfishery with one hand, while he slipped his other hand into the pocketbook of Canadian taxpayers for a bail-out of National Sea Products and the establishment of Fishery Products International. These two companies were subsequently given property rights to nearly one-half of all Atlantic Canada groundfish under the Enterprise Allocation program.

An excellent opportunity to effect capacity reduction by the prescribed 50 per cent had been lost by a government decision to rescue the offshore companies from insolvency at the expense of the federal treasury. The Liberal Government of the day had confessed that the saving of the offshore and the associated jobs was justification for the bail-out. The bottom line was that the larger number of poorly organized fishermen and plant workers involved in the traditional inshore fishery were seen as expendable. For many industry workers, their involvement with the industry was over.

The leaders of large companies, such as Henry Demone of National Sea Products, John Risley of Clearwater Fine Foods, or Vic Young of Fishery Products International, are confident that their corporations are well positioned to ride out the present crisis, and the future of their corporations is secure. They are now willing to suggest that in this time of crisis, the government ought to change the rules on assistance to a struggling industry. Vic Young, Chair of Fishery Products International, is already on record as being opposed to future bail-outs in *The Chronicle-Herald* of April 8, 1993. "We are dealing with a calamity of our own making," he said while addressing the Newfoundland Chamber of Commerce in April 1993. "It was caused by too much fishing, trying to create too many jobs and trying to sustain too many communities. We must accept responsibility for ensuring it never happens again." Young recommended the industry break free of government handouts once the stocks recover, suggesting that the fishery would be a mere shadow of its former importance to coastal communities by the time recovery had been achieved. "At the end of the process there will likely be as many as 15,000 people left out of the fishery," he speculated. Ironically, his company was one that was raised from the ashes of industry disaster by the Liberal rescue in the 1980s.

While directed financial assistance from government is admonished by the corporate upper crust, they have made it known they have a vision for the future. Their vision is not one of further financial buyouts, but one in which they are given free reign to accumulate an even greater share of groundfish quotas as independent fishermen succumb to the pressures and burdens of a prolonged slump in the fishery.

The Fisheries Council of Canada (FCC)—the trade association of fish and seafood producers and exporters in Atlantic Canada, and processors, distributors, and importers in Ontario—released its position paper on the future of the groundfishery at its annual meeting in New Brunswick in October 1994. The report, titled "A Vision for the Atlantic Fishery," called for abandonment of the present quota system and the longstanding fleet separation policy, established by the federal government in the late 1970s, prohibiting processors from owning and operating inshore fishing vessels. The fleet separation policy was designed to protect the interests of the inshore fishing sector from the capital-rich corporations intent on gaining greater control of the groundfishery. The "Vision" report suggested the processing sector (dominated by the large corporation offshore) should be allowed the option of purchasing boat quotas and owning and operating vessels which presently fish from the inshore portion of the groundfish TAC. Allowing processors the right to buy up inshore quotas and acquire inshore fishing enterprises would provide a more stable supply of raw product for these companies and would thus improve the economic prospects for the corporate sector. The proposed change of policy would also provide fishermen ready to leave the fishery with an increased probability of finding a buyer of their quota, easing a difficult transition and providing a market-driven method of harvesting capacity reduction.

Previous plans for capacity reduction of the inshore fleet had focused on a government-funded buyout of licences. The position put forward in the *Sou'Wester* on November 15, 1994, maintained that abandoning the federal fleet separation policy and allowing processors to buy inshore quotas would not necessarily mean that corporations would suddenly rush to control all access to supply of raw product. "It merely opens this up as one of several opportunities for securing raw material." The Fisheries Council of Canada (FCC) paper suggested the decision to sell or hold onto a licence and quota would remain with individual fishermen. Market forces rather than government intervention would determine whether widespread sale of quota would proceed.

An immediate response of qualified approval of the "Vision" report from Fisheries Minister Brian Tobin sent shivers down the spines of inshore fishermen. Tobin publicly announced that he liked what he had read of the "Vision," though cautioned that it was only one opinion of many he expected to hear. He later announced he would be convening a round table on the future of the Atlantic groundfish fishery in Halifax during March 1995, where "meaningful participation and input from all voices" within industry and government would be heard.

Reaction from the inshore was immediate and predictable. Various fishermen's groups condemned the FCC report, acknowledging that they must become better organized to counter the proposal. Gary Dedrick, acting director of the Southwest Nova Fixed Gear Association and Vice-President of the Scotia-Fundy region of the Eastern Fishermen's Federation, cautioned that the inshore fishery as it is known today would disappear within two years if the report was adopted by Ottawa. "Their intent is to push the inshore fisherman out. They want to own the fish, own the boats, own the plants, own the markets." To accept the processors' vision would mean that large corporations would control the entire groundfish resource by buying up quotas and fishing enterprises one by one, until they had a monopoly on the supply of raw material and control over market prices. Dedrick was quoted in the *Sou'Wester* on November 15, 1994, "What the report is saying is they want privatization and big company corporate ownership of the entire resource."

The federal government decided to proceed with establishment of regional harvesting adjustment boards less than one month after the release of the Fisheries Council of Canada (FCC) report. The senior level of government complained that the provinces, who were to partner implementation of Industry Renewal Boards, had been dragging their feet. Ottawa decided to set up four regional harvesting boards with the purpose of developing detailed plans to reduce capacity in the harvesting sector. The boards will then advise the federal government on licence and early retirement programs for fishermen. The programs feature voluntary measures designed to give fishermen incentive to leave the fishery. The harvesting adjustment boards are to complete their work before the end of 1995.

The federal government seemed committed to reducing the number of fishermen by half, but outspoken critics of the push by the FCC to free the restrictions on the processing sector cautioned that a simple reduction in numbers of fishermen would not be the most reasonable way to effect capacity reduction. Cliff Fanning, Executive Director of the Eastern Fishermen's Federation, made it clear that he and his membership were opposed to the direction of the federal government. He said compelling the small boat fishery to bear the brunt of harvesting capacity reduction would not address the real problem. He was quoted in the *Sou'Wester* on May 1, 1994: "The real problem is effort. We must take measures to reduce pressure on the resource, reduce the effort. This means looking at gear technology and re-allocating quota to the least destructive and labour-intensive fleets. If the capacity and the numbers of those dependent upon the fishery must be reduced it must be the decision of industry—not government, and especially not IRBs [Industry Renewal Boards] made up of government-appointed people."

While the federal government has never clearly indicated that it intends to accomplish harvesting capacity reduction on the backs of the inshore fixed gear sector as Fanning suggests, this is the overriding concern among small boat fishermen. John Leefe, Nova Scotia Progressive Conservative fisheries critic, recognizes the trepidation which has swept this sector and supports handline fishermen. His remarks appear in the *Sou'Wester* of May 1, 1994: "It's high time the Department of Fisheries and Oceans (DFO) and its Minister, Brian Tobin, recognized the vital importance of rebuilding our fishing around technologies which are, by their highly selective nature, agents of conservation.... DFO, while espousing conservation, continues to enforce rules which are not designed to rebuild a sustainable fishing industry. The department is intent on downsizing the number of participants in the industry, but instead may succeed only in creating a new, highly aggressive, less selective, more exclusive industry. Quite simply, the little guy is in danger of being squeezed out ... and nowhere is that more apparent than in the continuing restrictions imposed on handliners by DFO."

The issue of the rights of inshore small boat fishermen was taken up by Greenpeace Canada. Environmentalist groups had been slow to recognize the potential for ecological disaster but have recently joined forces with inshore fishermen to mount a strong voice in opposition to the destructive fishing practices employed by the offshore sector and mobile sector of the inshore. Bruno Marcocchio of Greenpeace Canada blasted the Cashin Report for blaming the inshore and warned against any government directives that might contribute to or effect a phasing out of the inshore hook and line sector, according to the *Sou'Wester* on January 1, 1994.

According to Marcocchio, "Rather than embarking upon a fundamental analysis of what has gone so terribly wrong with the fishery, the Cashin Task Force instead insists that the solution to the problem is the elimination of hordes of indolent unprofessional fishers. By reinforcing the Big Lie that the inshore is an inefficient social equalization plan we can no longer afford, the Task Force has designed a blueprint for the depopulation of rural Atlantic Canada. It fails to meet even the first goal set out in the terms of reference, which was to advise on the continued supply of the resource."

While Marcocchio articulated the general impression passed on to the public by the media, Cashin and the Task Force were later revealed to have hedged their argument to allow for an opposite interpretation. Under direct questioning as to the future of the inshore sector, Cashin disclosed during a public session held in Shelburne, Nova Scotia, in January, 1994, that the report includes a hidden obituary for the offshore. He referred to page forty of the report which states: "Capacity reduction should be based on the principle that coastal states would maintain priority access to resources upon which they have traditionally relied. For example, for northern cod there was a traditional inshore allowance (for vessels less than 65 ft.) of 115,000 tonnes. Principally, the harvesters of this were from along the northeast coast of Newfoundland and the coast of Labrador. It is unlikely that there will be a directed offshore fishery for northern cod in the future until the total allowable catch approaches or exceeds the traditional inshore allowance." In other words, those involved in the traditional inshore cod trap, gillnet, hook and line, and dragger fisheries would be the first re-entrants to the fishery when northern cod stocks off the northeast coast of Newfoundland and Labrador recovered sufficiently to justify a harvesting effort. This promise could be the ray of hope some fishermen have been looking for.

Acceptance of the Cashin Report's hidden agenda by the federal government would mean years, even decades, before the offshore would be allowed to harvest northern cod in the waters of northeast Newfoundland and Labrador, unless corporations persuaded Ottawa to abandon its fleet separation policy and allow processors to purchase inshore boat quotas and fishing enterprises. The Cashin Report made allowances for communities along the southern coast of Newfoundland which had served as the land base for offshore fishing efforts. According to the report, the offshore fleet would be allowed to pursue groundfish stocks on the southern Grand Banks, retaining use of trawler technology if it chose. "The same is true of other groundfish stocks including, for example, cod, flounder, and redfish on the southern Grand Banks, which have been the traditional resource base for communities on the south coast of Newfoundland and elsewhere. Offshore fleets and the communities dependent upon them should have priority access to these resources, whatever harvesting technology is used."

However, Irene Novaczek, Chair of the Canadian Oceans Caucus, points out that harvesting technology has very serious implications for long-term employment in the fishery. Novaczek was critical of the Cashin Report and mobile gear methods on CBC "Radio Noon" on December 21, 1993. "If you are going to allow bottom draggers and large trawlers to operate in the future then you have made a decision to go for an option that provides less employment; that allows a product of less quality to be brought ashore; which allows for ongoing impacts in terms of habitat destruction; which is non-selective; which is very dangerous in the wrong hands; and which is very difficult to control and monitor."

Supporters of the traditional hook and line method of harvesting groundfish point out that we have an opportunity during the moratorium, to carry out a fundamental analysis of our fishery and our reasons for fishing. Is it to provide dignified employment for Atlantic Canadians and to sustain coastal communities, or is it to provide as much profit as possible to a small number in the corporate sector? In recent years, the focus of the industry has slipped from dignified employment to profit, without participants consciously deciding to shift priorities. It is time for Canadians to reassess the direction we wish to take with the fishing industry. We owe a re-evaluation of the fishery and of our objectives in harvesting the ocean's resources to the industry, and to future generations who will have to contend with the consequences of our decisions.

The repeated failure of government adjustment and transition programs designed to assist former fishery workers dislocated by the crisis has made many aware that government quick fixes do not work and are not affordable. The Minister of Fisheries cannot be expected to rescue everyone in the fishery. Fishermen also realize they cannot rely on government to provide easy answers or to make the right decision to ensure a stable future. They have come through a period of mourning and now appear willing to help themselves. Fishermen and fishermen's representatives realize they must become actively involved in the process that will shape and hopefully preserve their way of life. They understand their communities will have to change. They also understand that any changes must not alter their coastal communities beyond recognition, but rather strengthen and rejuvenate them.

Chapter Ten

Running Out of Time: Blueprint for a Sustainable Groundfishery

As the present groundfish crisis inflicts an increasing toll on coastal communities, there does appear to be a general, though unpopular, resolution among most stakeholders in the Atlantic groundfishery to learn from the predicament. The prolonged closures now facing the Atlantic coast fishery will give everyone time to consider why the industry has reached this state of devastation. Those who have been associated with the fishing industry realize that historically this has been a sector subject to pronounced upturns and downturns. Perhaps it is also time to acknowlege that these cycles are determined by the way we approach the fishery. Even a casual examination of the history of the groundfishery reveals these crises are directly connected with overexploitation by the industry.

David Suzuki, Canadian biologist, environmentalist and broadcaster, maintains that overexploitation and the depletion of groundfish stocks are inevitable and the inescapable products of marriage between a prevalent exploitative mind-set and our increasing technological dexterity. Our rate of technological innovation and its application to the fishery have been staggering. We have allowed this harvesting power to rule and accelerate our dependence on the newest gadgetry. The resulting increase in fishing capability means that humans are now able to exploit marine resources as never before. According to the Environment and the Economy Symposium held on February 12, 1994, their sustainability is imminently threatened.

Are those in authority ready to accept Suzuki's analysis and the conclusions reached long ago by small boat fishermen that we must effect a fundamental change in the way we manage the fishery? The federal government must recognize the inherent dangers in the "Vision of the Atlantic Fisheries," offered by the corporate sector, that calls for the privatization of the groundfish industry and absorption of the inshore small boat fishery by the offshore sector. Will the sacrifices made now by coastal communities be in vain? In order to avoid repeating past mistakes, we need to look at the prospects for a revitalization of the marine environment off Canada's east coast.

Establishing a management system that recognizes the complexities and fragility of the marine environment and at the same time provides for a reasonable and sustainable harvest of the resources is slowly becoming a necessary priority, while a resource, an industry, and coastal communities of Atlantic Canada hang in the balance. Furthermore, the health of nearly all Atlantic groundfish stocks are declining, despite the imposed moratorium on fishing throughout most of the region and a reduction in fishing effort applied by foreign fleets just beyond the 200-mile limit. Scientists estimate that there has been a further decline in northern cod stocks of 90 per cent since the 1992 closure. The stock is teetering on the brink of extinction. Other Atlantic groundfish stocks are either showing signs of uncertain stability or are continuing to decline.

Studies have shown when stocks are fished down to a critical number, they tend to lose their genetic diversity. With a reduced gene pool, a stock is less likely to include those animals able to adapt to changes in environmental conditions, such as water temperature, salinity, dissolved oxygen, or a change in their physical environment, such as increased suspension of sediments in the water column or pollution. There is a nagging fear that northern cod is no longer equipped genetically to respond to the pressures of a harsh environment. Our overfishing has critically reduced the gene pool of northern cod and thus the inherent ability of the stock to adapt to changing conditions, rendering it less capable of rejuvenation.

A recent study conducted by Myers et al. and published in *Science* magazine dated August 25, 1995, offers some hope that groundfish stocks are more resilient than suspected. It showed that groundfish driven to low population levels may be able to overcome difficulties associated with reduced genetic diversity and even some of the pressure applied by a harsh environment if commercial fishing is stopped and stock decline is arrested. The study showed that despite a dramatic drop in the abundance of spawning fish, among 125 major commercial species studied, Labrador northern cod, Georges Bank haddock, and Georges Bank silver hake showed evidence of "increased survival at lower population levels." The researchers concluded that "the effects of overfishing are, at this point, still generally reversible."

This is the hope that many in the industry have been clinging to, trusting that despite the crash of groundfish stocks, conditions will improve. Most cannot contemplate a world without fish and, therefore, hold on to their licences, boats, and gear while hoping for a return of the stocks. They dream of a sustainable fishery that will meet the needs of coastal communities.

"Sustainable development" is one of the newest phrases being floated these days. The term is often criticized as an oxymoron and, therefore, carries conflicting definitions. A report titled "Our Common Future: Report of the World Commission on Environment and Development" by Brundtland defined sustainable development as "development that meets the needs of the present without compromising the ability of future generations to meet their own needs." While the general principle of sustainable development of the fisheries is easy for most to support, it is far more difficult to agree on what it would mean when applied to the Atlantic coast groundfishery. For those concerned with the integrity of the ocean's biological system, sustainable development is often used interchangeably with ecologically sustainable development. To a stakeholder concerned about the survival of a coastal community or to a big business executive, sustainable development means different things. For the former, sustainable development would mean little significant interference with the natural system and promotion of passive and selective methods of fishing. For those concerned with the long-term survival of fisheries dependent coastal communities, sustainable development is used interchangeably with economic sustainable development, stressing maintenance of a catch level recorded for an area over time and that level of catch necessary for the continued support of the local coastal economy.

For the corporate executive there would be less concern for the sustainability of the local fishery resource and more concern for the next business opportunity. If a president of a large fish company were to determine that there were $100 million of fish in the ocean available for exploitation, then company policy would lean toward catching every one of those fish at the least cost, perhaps with a view to ploughing the profits into another venture, such as a chicken farm in Idaho should fish stocks fail. Under the fisheries theory of economics practised by large profit-driven companies, the bottom line is often all that matters and this may justify at the corporate level the prosecution of vulnerable stocks to the detriment of future generations and the sustainability of the groundfish resource. While the capital of large corporations remains mobile, fishermen are anchored, having made risky personal investments in boats, gear, vehicles, and have often taken out mortgages on their homes. They will sink or swim, depending on the abundance of local stocks and an adequate market price. Corporate capital, on the other hand, is very mobile, shifting to wherever there is a profit to be made. The fortunes of the large company sector are therefore loosened from the constraints of sustainable development and are free to pursue profit at any cost. Moving into coastal communities, corporate capital upsets the local economy; moving to the next business opportunity, corporate capital leaves workers displaced and resources pilfered.

This scenario of resource depletion and abandonment of local communities marks the present crisis in the Atlantic coast groundfishery. The large offshore companies, faced with shortages of locally caught groundfish, have closed plant after plant in an effort to rationalize their operations and have moved to purchase raw product on the world market. They have successfully adjusted to the hardships created by resource depletion and have shown they are not disabled when the sustainability of stocks is undermined. National Sea, quickly followed by Fishery Products International, and later Seafreez Foods Limited in 1987, each began to purchase frozen at sea, headed and gutted haddock and other groundfish from foreign interests. The purchases were originally made to supplement gaps in the supply of raw material left by dwindling domestic fish quotas. The purchased catches from foreign vessels now constitute a major portion of the production in the few offshore plants still operating. With an external supply and an aging fleet ready for decommissioning, the next move was obvious.

The large corporations began selling off or offering for salvage their old trawlers in 1991. Fishery Products International is well along the road to disposing of its fleet. While it still operates twenty-four trawlers, six scallop vessels, and a shrimp freezer trawler, the company offered for sale twenty-five of its stern trawlers in 1993. Sales progressed well with thirteen sold by April 1994. Most were destined for South America and other warmer climes. National Sea Products has sold some of its fleet and early in 1994 sold the factory freezer trawler *Cape North,* a vessel whose efficiency of operation depended on a high volume fishery. Seafreez Foods Limited has also cut its fleet, choosing to sell four trawlers acquired from National Sea when Seafreez purchased the Burgeo plant in 1990.

Despite the Atlantic groundfish closures, the large offshore processing companies have been able to procure sufficient product to keep established world markets open. And with the sale by large portions of their trawler fleet, these giants have turned red ink into black. For National Sea Products, 1994 was its first profitable year after a string of losses dating back to 1987. Also, while they were able to make a profit on groundfish purchased from Russia and China, the big companies diversified, promoting new value-added products such as battered foods that required additional processing. National Sea has begun processing battered chicken as well as fish, increasing the likelihood that a chicken farm or two may be in the company's future. Explaining the shift in emphasis away from fish products, National Sea Product's President, Henry Demone, told *The Chronicle-Herald* on August 11, 1994, "The outlook for fishing is gloomy, but we're hardly a fishing company anymore.... We're basically a frozen food processing company that imports raw material from around the world," one which is less and less dependent upon the sustainability of the northwest Atlantic groundfishery.

Research has shown that the pursuit of economically sustainable development of the groundfishery does not coincide with ecologically sustainable development of the resource. According to Ludwig et al. in an article from *Science,* "...Resources are inevitably overexploited, often to the point of collapse or extinction." The goal of economic sustainability is virtually impossible to achieve. With the world's population doubling every forty years, it becomes obvious that we will require twice as many fish in forty years as today. The ocean's ecosystem cannot withstand the pressure as we have already discovered. No measure of ocean production enhancement will succeed in meeting the requirements of an ever expanding human population. For this reason, economic sustainability must be made subordinate to ecologically sustainable development. Our approach to the Atlantic groundfishery cannot continue to be dictated by an emphasis on growth and short-term profit. Instead, we have to decide how large our fishery can be, the most appropriate method of harvest, and how much we will be able to rely on the wild stock of groundfish in the future.

Fisheries Minister Brian Tobin and the DFO have recently made a commitment to rebuild groundfish stocks and to exercise caution when setting quotas and determining when it is appropriate to open closed fisheries. What remains to be seen is which sector(s) will be able to ride out the storm until stocks recover sufficiently to reopen the groundfishery. The shape of the industry through "industry renewal" by the federal government is underway. Will Ottawa side with the corporate sector that is pushing for privatization and access to inshore quota, or will the federal government implement a process of renewal that favours energy efficient and ecologically sound harvesting technologies and fishing practices? To provide maximum social and economic benefit for Atlantic Canada and to better ensure resource and ecosystem sustainability, the industry needs a reallocation of harvesting rights to favour the small boat hook and line sector.

Yet, more than a restoration of stocks is needed and more than a reallocation of harvesting rights is required. Responsibility for management of the groundfishery must be placed in the hands of those who are most dependent upon it for their livelihoods—fishermen. Bonafide independent inshore fishermen should be given the right to co-manage the fishery, the same sector which should be given priority rights to a rejuvenated resource and the sector which can be expected to live by regulations they have worked to develop. Management of the groundfishery should take place through a partnership between DFO and inshore fishermen, with the latter helping to control policies and regulations and to determine levels of catch. For sustainable development of the groundfishery to be realized, more than economic and ecological concerns must be addressed. Stakeholder representation and input must be considered.

The need to decentralize and democratize the resource management process was recognized by the Brundtland Commission in "Our Common Future: Report of the World Commission on Environment and Development." "What is needed most is that the population have the knowledge and the will to follow up, which means the population being allowed to participate in the decisions which will affect the environment. This is best achieved by decentralization of control of the resources which the local population is dependent on, and by providing the population with real input into how these resources should be used."

The DFO has recently broached the subject of co-management of the resource with industry, admitting that it can no longer continue with a "patchwork quilt" of fisheries management regulations that are inefficient and expensive to administer. However, it has not indicated how it intends to pursue the matter and has not indicated which industry players will be invited to participate. While government may struggle to find a formula for co-management of the fisheries, the private sector has already come forward with a proposal worth serious consideration. In a paper titled "Managing the Atlantic Groundfishery, Formula for Change," authors Geoffrey Hurley, of Hurley Marine Consulting of Dartmouth, and David Gray, an operational research scientist at St. Mary's University, Halifax, have devised a formula for co-management which they suggest is "consistent with the essential characteristics of an ideal fishery," including: "biologically and economically sustainable; easily monitored, controlled and enforced; and uses selective fishing gear that does not destroy bottom habitat." The functional body under the proposal is called a "Community Management Committee, CMC" and would be responsible for developing a fishing plan for each community subscribing to the co-management process. Membership on the CMC would include fishermen, area fish processors, seafood distributors, and other stakeholders. Representatives of the CMCs would also have a seat on a "Regional Community Management Committee (RCMC)" which would be responsible for determining regional policies and resolving disputes. The RCMC would also work closely with the Fisheries Resource Conservation Council (FRCC) which, in turn, is designed to be an intermediary for government and industry. The Department of Fisheries and Oceans would be responsible for enforcing compliance with the fishing plan and providing technical advice to FRCC and CMCs on "conservation-related matters, such as monitoring the biological condition of stocks, determining long-term sustainable yields for various fisheries and developing selective and environmentally friendly fishing gear."

The key function of a co-management system must be harvesting capacity reduction. Not because there are "too many fishermen chasing too few fish," but because overfishing is largely a result of the use of modern technology and presently there is too much harvesting capacity (powerful draggers and trawlers) tied to Atlantic Canadian wharves awaiting the opportunity to fish again. Some dragger fishermen would welcome the chance to switch to the longlining or hook and line method if that is provided as an option. The incentive to do so would be greater if the benefits of increased quota, afforded by recovery and growth of stocks, are assigned to those fishing with fixed gear.

The federal government should cover the costs of retiring dragger licences should the captains of these enterprises decide they wish to leave the fishery. A one-time buy-back offer should be made to remove these fishermen from the industry. By requiring retirement of the dragger enterprise, a repeat of the bail-out of the offshore corporate sector in the early 1980s would be avoided.

However, government should not be obligated to assist in retirement of offshore trawlers from the fishery with the large corporations effectively relinquishing any claim to compensation by buying into the Enterprise Allocation program instituted in 1982. For the retirement of inshore draggers some equitable formula could be determined to account for the private investment already made and perhaps to give consideration to the loss of income opportunity. For those willing to convert their vessels, funding should be made available for the refit. One should not anticipate that all or even the majority of dragger fishermen would want to convert to the more labour-intensive hook and line method, though that possibility exists.

The removal of excess harvesting capacity from the industry by retirement of licences and conversion of draggers to a fixed gear fishery, will take away the major impediment to recovery of stock health. The longline and handline sectors should not be subject to quota management control. Rather, they could be placed under an allowance system, where the only restraints would be the minimum size of hooks used and the number of hooks fished.

Other methods could fit the fishery model established under co-management. For example, gillnets are not inherently destructive, only potentially so, and the danger posed by this method is in direct proportion to the carelessness of its user. Gillnet fishermen would be allowed to ply their trade, as long as the areas fished and the number of nets employed are restricted and rigidly enforced. Gillnets should not be fished on Georges Bank, in the Bay of Fundy, or any other area where strong tides occur because of the increased chance the gear may be lost. Nets must remain attached to the vessel at all times and be constructed either entirely or partially of biodegradable material to reduce the harmful effect of ghost fishing. Similarly, the traditional cod trap fishery of Newfoundland and Labrador should be given priority access to the northern cod stocks once recovery is sufficiently underway. However, this sector must move to a larger mesh size of 89 mm, or greater, to minimize the amount of small fish taken.

For a thorough recovery of the groundfishery, measures to reduce harvesting capacity and to eliminate technologies which destroy bottom habitat must be supplemented by steps to protect the feed stocks of groundfish. Fish need to eat in order to grow, so their feed stocks must be returned to health as well. Protection of feed stocks is a crucial part of the recommended ecological approach to the fishery. Reduced pressure on these stocks will return dividends later as groundfish feed on their favourite food, grow more quickly and to greater size, and add to stock size. Those involved in harvesting and processing these feed stocks will also benefit as herring, mackerel, capelin, and other stocks grow to a point at which a carefully monitored commercial catch is again permissible.

Another ecologically sound measure to support a sustainable fishery is the establishment of protected spawning and nursery areas. The creation of these no-harvest zones have had strong industry support from all components of the inshore sector and have resulted in areas such as Browns and Georges banks, off southwestern Nova Scotia, being closed to fishing during the critical spring spawning period for haddock and cod. Although the federal government has complied with the interests of inshore fishermen in extending these closures, it also denies the protected areas have any intrinsic biological value. To admit that these closures increase the reproductive success rate of groundfish would leave federal bureaucrats open to attack for not extending the closures to include other sensitive spawning and nursing areas.

More and more interest is being shown in aquaculture—the cultivation of groundfish in a controlled marine environment—as an opportunity for those attached to the sea to maintain their way of life. Some suggest that aquaculture could be a bridge to groundfish stocks returning to health. Aquaculture requires a high level of skill and keen business sense. Fishermen attracted to this alternative will require assistance with the transition. The culture of farm-raised fish will help fishermen to maintain contact with the sea and perhaps help the industry to hold market contacts which might otherwise be lost should free swimming stocks continue their decline.

With much of Atlantic Canada under a groundfish moratorium, there is a very real danger that if stocks return, markets for our fish products may be lost. Unique to this latest groundfish crisis is the unbelievable change markets have already undergone. It used to be that if Canadian cod landings would dip, then prices would rise as an immediate response. Instead, the market has already shed its dependence on the supply of Canadian groundfish. Those needs are now met by the harvest of different stocks of cod and groundfish on the world's oceans halfway around the globe, and the shipping of this less expensive product fished by huge factory freezer trawlers has flooded the markets, driving the price of groundfish down. Almost any Atlantic Canada processor today would admit he has no idea how he could sell cod if a boat were off-loaded at his plant. It is now obvious that the pressures of the global marketplace will have a significant role in determining the success of any recovery in the groundfishery. Our harvesting and processing sectors must realize that even with a return of stocks, the changes in the marketplace dictate that there will not be a return to the status quo. The Canadian industry must now begin carving a new share of the market, one which rewards the harvest of a quality raw product and the production of a quality end product.

Canadians may take solace in the Norwegian example of recovery. They have already suffered a similar decline in stocks and similar threats to their markets, but they are now on their way to recovery. Norwegian fishermen have had a history of close involvement with government in the design and management of their groundfishery. They, too, suffered a collapse of fish stocks in the 1970s and again in the mid-1980s. Scientists had optimistically predicted cod stocks would yield a catch of 900,000 tonnes in 1990. The reality was that overfishing and the fishing of feed stocks had decimated the cod, prompting quota cuts to 113,000 tonnes in 1990, and fear that a closure of the fishery was imminent. Tough measures were taken to avoid closure and assist stock recovery. One measure involved the elimination of purse seining for capelin, the primary food source of cod. In addition, large trawlers were restricted to waters beyond 19 km (12 mi.) from the coast, while small draggers were excluded from waters 6 km (4 mi.) from shore. The Norwegian industry expressed concern for seafloor damage from otter trawlers by replacing most bottom trawlers with midwater trawlers. A total ban was placed on discards and reports of a high percentage of small fish being caught in one area prompted immediate closure by the authorities. The Norwegian government had enough foresight in the 1970s to begin to invest in the development of aquaculture as a backup to the traditional fishery and as a supplement to help satisfy established markets for fish products.

We can also look to other coastal nations which have underestimated the increasing ability of their fishing fleets to deplete the marine resources. The decimation of the Atlantic groundfish stocks should have been recognized as inevitable. We are now paying a terrible price for our nearsightedness. We failed to realize that nature does not expand its storehouse of resources simply to supply our needs and gratify our greed.

While the groundfishery continues at a virtual standstill, we have a unique opportunity to properly restructure the way we harvest groundfish. If the federal government were to commission a public inquiry to investigate the causes of the groundfishery collapse and to seek input to determine the face of a future fishery, industry workers would have an opportunity to get involved at a grassroots level. This type of inquiry and a commitment to follow its recommendations are critical if we wish to provide for the future of the groundfishery and the future of hundreds of Atlantic Canada coastal communities. It would be necessary for the inquiry to restructure management so that DFO has a reduced role and the fishing industry assumes co-management. In addition, the inquiry should determine which fishing gears and practices are most selective and environmentally friendly and make recommendations on a program to convert draggers and trawlers to more sustainable harvesting technologies. Finally, it is critical that we set the fishery on a course which will firstly provide ecological sustainability of the resource and secondly provide for economic sustainability of coastal communities.

There is some indication that the forces of nature in combination with government measures taken thus far, such as closures and quota reductions, may have arrested the decline in stocks. Scientists have recently reported finding schools of young cod off Newfoundland, and fishermen fishing waters off southern Nova Scotia insist that groundfish are now as plentiful as they were in 1987. There are glimpses of hope in what may seem overwhelming despair, though a conservative approach is still in order. We must make caution, reason, resolution, and hope our watchwords as we work to ensure a return to a sustainable fishery and a cherished way of life, with a new commitment to conservation.

Bibliography

"A Draft Brief Prepared for Municipal Units in Digby, Yarmouth and Shelburne Counties, To be Presented to the House of Commons Standing Committee on the Fisheries." January 20, 1990.

Angel, J. R.; Burke, D. L.; O'Boyle R. N.; Peacock, F. G.; Sinclair, M.; and Zwanenburg, K. C. T. 1994. "Report of the Workshop on Scotia-Fundy Groundfish Management from 1977 to 1993," *Canadian Technical Report of Fisheries and Aquatic Sciences.* no. 1979. Dartmouth.

"Annual Statistical Review of Canadian Fisheries Marketing Services Branch." 1968. Fisheries and Marine Services. Ottawa: Environment Canada.

Atlantic Fisherman. March, 1993. Halifax.

Barrett, Gene and Apostle, Richard, eds. 1992. *Emptying Their Nets: Small Capital and Rural Industrialization in the Nova Scotia Fishing Industry.* Toronto: Toronto Unviversity Press.

Barrett, Gene and Davis, Anthony. 1983. "Floundering in Troubled Waters." Occasional Paper no. 1-0783. St. Mary's University, Halifax.

Beamish, F. W. H. 1966. "Muscular Fatigue and Mortality in Haddock, *Melanogrammus aeglefinus,* Caught by Otter Trawl." *Fisheries Research Board of Canada.* vol. 23, no. 10.

Beck, F. A. Van; Leeuwen, P. I. Van; and Rijnsdorp, A. D. 1990. *Netherlands Journal of Sea Research.* vol. 26. no. 1.

Bjordal, A.; Laevaster, T. 1990. "Effects of Trawling and Longlining on the Yield and Biomass of Cod Stocks—Numerically Simulated." ICES. CM 1990/G:32.

Brawn, Vivian. 1961. "Reproductive Behaviour of the Cod (*Gedus Callarias L.*)." *Behaviour.* vol. 28.

Brundtland, G. H., ed. 1987. "Our Common Future: Report of the World Commission on Environment and Development." Oxford: Oxford University Press.

Caddy, J. F. 1973. "Underwater Observations on Tracks of Dredges and Trawls and Some Effect of Dredging on a Scallop Ground." *Journal of Fisheries Research Board of Canada.* vol. 30.

Cameron, Silver Donald. April/May, 1990. "Net Losses—The Sorry State of our Atlantic Fishery." *Canadian Geographic.* Vanier, Ontario.

Canover, Robert J.; Wilson, Scott; Harding, Gareth C. H.; and Vass, W. Peter. 1994. "Climate, Copepods and Cod: Some Thoughts on the Long-Range Prospects for a Sustainable Northern Cod Fishery." *Proceedings of Coastal Zone Canada '94.* vol. 4. 1732. Halifax.

Cashin, Richard. 1993. "Charting A New Course: Towards the Fishery of the Future. Task Force on Incomes and Adjustments in the Atlantic Fishery." (Cashin Report.) Ottawa: Minister of Supply and Services Canada.

CBC "Radio Noon." December 21, 1993. Halifax.

Chantraine, Pol. 1993. *The Last Cod-Fish: Life and Death of the Newfoundland Way of Life.* Montreal: Robert Davies Publishing.

Chong, K. C.; Dwipongga, A.; Ilyas, S.; and Martosubrota, P. 1987. "Some Experience and Highlights of the Indonesian Trawl Ban Bioeconomics and Socioeconomics." *Indo-Pacific Fisheries Commission RAPA Report 1987.* vol. 10.

Bibliography

Dalhousie News. April 26, 1995. Halifax: Dalhousie University.

De Groot, S. J. 1984. "The Impact of Bottom Trawling on the Benthic Fauna of the North Sea." Ocean Management. vol. 9. p. 177-190.

Department of Fisheries and Oceans Press Release NR-HQ-85-07BE. December 8, 1985.

Department of Fisheries and Oceans Press Release NS-MQ-88-072E, December 30, 1988.

Department of Fisheries and Oceans Press Release NR-HQ-89-009E. March 20, 1989.

Department of Fisheries and Oceans Press Release NR-HQ-89-025E. July 12, 1989. "Action Plan on Overcapacity."

Department of Fisheries and Oceans Press Release. B-SF-90-11E. February 20, 1990. Halifax: Communications Branch Department of Fisheries and Oceans.

Department of Fisheries and Oceans Press Release, DF-NR-90-13E. May 10, 1990. "130 Square Mesh Codends Fall Short of Task Force Target."

Department of Fisheries and Oceans Press Release, NR-SF-93-01E. January 11, 1993.

Department of Fisheries and Oceans. "The Science of Cod." *Fo'c'sle.* vol. 8. no.2. Minister of Supply and Services Canada.

Dunne, E.B. 1990. "Report of the Implementation Task Force on Northern Cod." (Dunne Report) *Communications Directorate.* Ottawa: Department of Fisheries and Oceans.

Environment and the Economy Symposium. February 12, 1994. Dalhousie University, Halifax.

"Fisheries Industry Profile and Impact Study." 1973. Halifax: Nova Scotia Department of Development.

Fishing News International. February and August 1993. London, England. vol. 32. nos. 2 and 8.

Graham, M. 1955. "Effect of Trawling on Animals in the Sea Bed." *Deep Sea Research 3, Supplement.*

Hache, Jean-Eudes. 1989. "Report of the Scotia-Fundy Groundfish Task Force" (Hache Report). Ottawa: Department of Fisheries and Oceans.

Hansard, House of Commons. November 8, 1985. Ottawa.

Hansard, House of Commons, October 2, 1989. Ottawa.

Hansard, House of Commons. January 22, 1990. Ottawa.

Hansard, House of Commons. April 10, 1990. "Minutes of the Proceedings and Evidence of the Standing Committee on Forestry and Fishery. Ottawa."

Hansard, House of Commons. October 2, 1990. Ottawa.

Hansard, Nova Scotia House of Assembly, Committee on Resources. April 24, 1990.

Harris, Leslie. 1990. "The Independent Review of the State of the Northern Cod Stock." (Harris Report.) Communications Directorate. Ottawa: Department of Fisheries and Oceans.

Hawkins, A. D.; Chapman, K. J.; and Symonds, D. J. August 26, 1967 "Spawning of Haddock in Captivity." *Nature.* vol. 215.

Holme, N. A. 1983. "Fluctuations in the Benthos of the Western English Channel." *Oceanologica Acta.* no. SP.

Hurley, Geoffrey V.; and Gray, David F. 1994."Managing the Atlantic Groundfish Fishery." *Fisheries Research Board of Canada.* vol. 19, no. 3.

Jenner, K.; Strong, K. W.; and Pocklington, P. 1991. "A Review of Fisheries Related Sea Bed Disturbances in the Scotia-Fundy Region." *Report to the Industry, Services and Native Fisheries Branch, Department of Fisheries* and Oceans. no. 166.

Keats, D.; Steele, D. H.; and Green, J. M. 1986. "A Review of the Recent Status of the Northern Cod Stock (NAFO Divisions 2J, 3K and 3L) and the Declining Inshore Fishery."

Ketchen, K. S. 1947. "An Investigation Into the Destruction of Grounds by Otter Trawling Gear." *Progress Report Fisheries Research Board. of Canada.* vol. 73.

Kirby, Michael J. L. 1982. "Navigating Troubled Waters—A New Policy for the Atlantic Fisheries." (Kirby Report.) Ottawa: Minister of Supply and Services.

Lamson, Cynthia and Hanson, Arthur J., eds. 1984. "Atlantic Fisheries and Coastal Communities: Fisheries Decision Making Case Studies." Dalhousie Ocean Studies Program. Halifax.

Ludwig, Donald; Hilborn, Ray; and Walters, Carl. April 2, 1993. "Uncertainty, Resource Exploitation, and Conservation: Lessons from History." *Science.* vol. 260.

Main, J. and Sangster, G. I. 1988. "A Progress Report on an Investigation to Assess The Scale Damage and Survival of Young Gadoid Fish Escaping From the Codend of a Demersal Trawl." *Scottish Fisheries Work Paper.* no. 3188.

"Maritime Fishermen's Union Against Proposed Capelin Fishery." *Sou'Wester.* March 15, 1994.

Martin, Cabot. 1992. *No Fish and Our Lives: Some Survival Notes for Newfoundland.* St. John's: Creative Publishers.

"Minutes of Proceedings and Evidence of the Standing Committee on Forestry and Fisheries." April 10, 1990. Ottawa.

Mitchell, C. C. and Frick, H. C. 1970. "Government Programs of Assistance for Fishing Craft Construction in Canada: An Economic Appraisal." *Canadian Fisheries Reports.* No. 14. Ottawa: Economics Branch, Fisheries Service, Department of Fisheries and Forestry. Information Canada.

Myers, R. A.; Barrowman, N. J.; Hatchings, J. A.; and Rosenberg, A. A. August 25, 1995. "Population Dynamics of Exploited Fish Stocks at Low Population Levels." *Science.* vol. 269. no. 5227.

National Fishermen. 1991. Yearbook. Rockland, Maine. vol. 71. no. 13.

Neilson, J. D.; Waiwood, K. G.; and Smith, S. I. 1989. "Survival of the Atlantic Halibut (*Hippoglossus hippoglossus*) Caught by Longline and Otter Trawl Gear." *Canadian Journal of Fisheries Aquatic Science.* vol. 46.

"Northern Cod: A Fisheries Success Story." 1980. Ottawa: Communication Branch Department of Fisheries and Oceans.

O'Boyle, R. N.; Sinclair, A. F.; and Hurley, P. C. F. 1991. "A Bioeconomic Model of an Age-Structured Groundfish Resource Exploited by a Multi-Gear Fishing Fleet." *Marine Science Symposium.* International Council for the Explorattion of the Sea (ICES). vol. 193.

"Policy for Canada's Commercial Fisheries." 1976. Ottawa: Department of Environment, Fisheries and Marine Services.

Presentation by Kingsley Brown, Fishing Masters Association, to the Hache Task Force. August 30, 1989. Halifax.

Pross, A. Paul. 1987. "Economic Resurgence and the Constitutional Agenda: The Case of the East Coast Fisheries." *Queen's Studies on the Future of the Canadian Communities.* Institute of Intergovernmental Relations. Queens University, Kingston.

Radovich, J. 1981. "Resource Management and Environmental Uncertainty." *Lessons from Coastal Upwelling Fisheries.* M. R. Glantz and J. D. Thompson, eds. New York: John Wiley & Sons.

Rapporteur's Report. September 23-25, 1980. Gulf Groundfish Seminar. Memramcook.

Reise, K. 1982. "Long-term Changes in the Macrobenthic Invertebrate Fauna of the Wadden Sea: Are Polychaetes About to take Over?" *Netherlands Journal of Sea Research.* vol. 16.

"Report of Fisheries Resource Conservation Council to Minister Brian Tobin. Re: Science Priorities." January 28, 1994.
"Resource Prospects for Canada's Atlantic Fisheries 1980-1985." 1980. Ottawa: Supply and Services Canada.
Rose, George A. December 2, 1993. "Cod Spawning on a Migration Highway in the North-west Atlantic." *Nature.* vol. 366.
Sahrhage, Dietrich and Lundbeck, Johannes. 1992. *A History of Fishing.* Berlin: Springer-Verlag.
Sars, G. O. 1876. "On the Spawning and Development of the Codfish." *Report U.S. Fish Commission for 1873-1875.* no.3.
Saxon, F. 1980. "Coral Loss Could Deplete Fish Stocks." *Catch.* vol. 7(8).
"Shifting Down." January/ February 1989. *Atlantic Business.* Halifax.
Soldal, A. V.; Engas, A.; and Isaken, B. 1993. "International Council for the Exploration of the Sea." *Marine Science Symposium.* vol. 196.
Sou'Wester. May 15, 1989. Yarmouth.
Sou'Wester. September 1, 1989.
Sou'Wester. January 15, 1990.
Sou'Wester. February 1, 1990.
Sou'Wester. March 1, 1990.
Sou'Wester. November 1, 1993.
Sou'Wester. January 1, 1994.
Sou'Wester. April 15, 1994.
Sou'Wester. May 1, 1994.
Sou'Wester. November 15, 1994.
Sou'Wester. July 1, 1995.
Sou'Wester. July 15, 1995.
Sou'Wester. August 15, 1995.
Submission and Response of the Nova Scotia Dragger Fishermen's Association, to be considered by Division 4X and 5 Groundfish Seminar. Presented May 26, 1986, Yarmouth.
Suuroven, P.; Lehtoven, E.; Tschernij, V.; and Orrensalo, A. 1993. "Survival of Baltic Herring (*Clupea harengus* L.) Escaping from a Trawl Codend and Through a Rigid Sorting Grid." International Council for the Exploration of the Sea (ICES) Statutory Meeting, Fish Capture Committee. CM 1993/B:14.
"Ten Years of Fisheries Development, 1965-1975." 1975. Environment Canada, Fisheries and Marine Service.
The Chronicle-Herald. October 5, 1989. Halifax.
The Chronicle-Herald. March 10, 1990.
The Chronicle-Herald. April 23, 1990.
The Chronicle-Herald. June 21, 1990.
The Chronicle-Herald. June 18, 1991.
The Chronicle-Herald. July 4, 1991.
The Chronicle-Herald. July 19, 1991.
The Chronicle-Herald. October 24, 1991.
The Chronicle-Herald. November 26, 1992.
The Chronicle-Herald. April 8, 1993.
The Chronicle-Herald. April 15, 1993.
The Chronicle-Herald. June 21, 1993.
The Chronicle-Herald. July 20, 1993.
The Chronicle-Herald. November 8, 1993.
The Chronicle-Herald. July 25, 1994.

The Chronicle-Herald. August 11, 1994.
The Chronicle-Herald. January 26, 1995.
The Chronicle-Herald. February 2, 1995.
The Chronicle-Herald. September 6, 1995.
The Coast Guard. February 1, 1994. Shelburne.
The Coast Guard. August 22, 1995.
The Daily News. December 21, 1993. Halifax.
The Guardian Newspaper. March 8, 1988. Clark's Harbour.
The Guardian Newspaper. May 17, 1988.
The Guardian Newspaper. October 4, 1988.
The Guardian Newspaper. December 20, 1988.
The Guardian Newspaper. February 21, 1989.
The Guardian Newspaper. May 16, 1989.
The Guardian Newspaper. October 31, 1989.
The Guardian Newspaper. January 9, 1990.
The Guardian Newspaper. May 15, 1990.

Watt, John W. 1963. "A Brief Review of the Fisheries of Nova Scotia." Halifax: Nova Scotia Department of Trade and Industry.

Wells, Kennedy. 1986. *The Fishery of Prince Edward Island.* Charlottetown: Ragweed Press.

Whittington M. and Williams G. 1980. "Pressure Groups: Talking Chameleons." *Canada in the 1980s.* Toronto: Methuen.

World Resources Institute. 1992. *A Report of the World Resources Institute in Collaboration with the United Nations Environment Program and the United Nations Development Program.* New York: Oxford University Press.

Zaferman, M. L.; and Serebrov, L. I. 1989. "On Fish Injured When Escaping Through The Trawl Mesh." International Council for the Exploration of the Sea (ICES). CM/B:18.